ACCLAIM FOR Survival of the Prettiest

"[A] sprightly, spunky, well-written treatise on the Darwinian science of looking good."
—*Entertainment Weekly*

"[Etcoff's] writing is a spirited blend of scientific analysis and cultural observation, mixing findings from cutting-edge social research, passages from literature and gossipy tidbits worthy of *People* magazine."
—*Minneapolis Star Tribune*

"In *Survival of the Prettiest*, Ms. Etcoff digs through a mountain of scientific research, anecdotal illustrations and sharp observations to find the true meaning of beauty."
—*Washington Times*

"By drawing widely from anthropological, psychological, biological, and archeological literature, Etcoff discerns surprising similarities in the ways humans have responded to beauty across diverse cultures throughout the millennia."
—*Publishers Weekly*

"*Survival of the Prettiest* synthesizes much recent research and will convince all but the most ideologically rigid that sensitivity to beauty is to a very large extent hard-wired in the human brain."
—*The Wall Street Journal*

NANCY ETCOFF

Survival of the Prettiest

Nancy Etcoff has an M.Ed. from Harvard, a Ph.D. in psychology from Boston University, and has held a post-doctoral fellowship in brain and cognitive sciences at MIT. She is currently a faculty member at Harvard Medical School and a practicing psychologist at Massachusetts General Hospital. She lives in Cambridge, Massachusetts.

Survival of the Prettiest

THE SCIENCE OF BEAUTY

Nancy Etcoff

ANCHOR BOOKS
A DIVISION OF RANDOM HOUSE, INC.
NEW YORK

To My Mother and To
the Memory of My Father

FIRST ANCHOR BOOKS EDITION, JULY 2000

Copyright © 1999 by Nancy Etcoff

All rights reserved under International and Pan-American
Copyright Conventions. Published in the United States by Anchor Books,
a division of Random House, Inc., New York, and simultaneously in Canada by
Random House of Canada Limited, Toronto. Originally published in
hardcover in the United States by Doubleday, a division of
Random House, Inc., New York, in 1999.

Anchor Books and colophon are registered
trademarks of Random House, Inc.

The Library of Congress has cataloged the Doubleday edition as follows:
Etcoff, Nancy L., 1955–
     Survival of the prettiest : the science of beauty / Nancy Etcoff.
          p.   cm.
     Includes bibliographical references and index.
     1. Beauty, Personal—Social aspects.   2. Sexual attraction.
3. Natural selection.   I. Title.
  GT499.E85   1999
391.6—dc21   98-41332
CIP

Anchor ISBN 0-385-47942-5

Author photograph by Ken Schles
Book design by Gretchen Achilles

www.anchorbooks.com

Printed in the United States of America

10   9   8   7   6   5   4   3   2   1

Contents

Survival of the Prettiest

CHAPTER ONE

Introduction: The Nature of Beauty

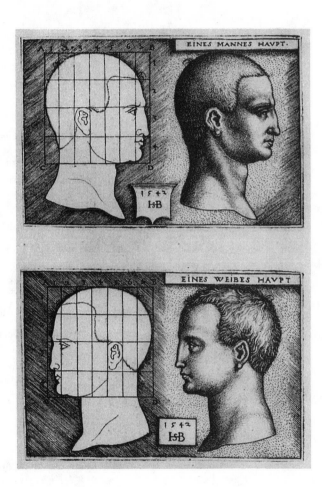

The three wishes of every man: to be healthy, to be rich by honest means, and to be beautiful.

— PLATO

There must . . . be in our very nature a very radical and wide-spread tendency to observe beauty, and to value it. No account of the principles of the mind can be at all adequate that passes over so conspicuous a faculty.

— GEORGE SANTAYANA

(Yes, I know. You haven't the slightest idea what I'm talking about. Beauty has long since disappeared. It has slipped beneath the surface of the noise, the noise of words, sunk deep as Atlantis. The only thing left of it is the word, whose meaning loses clarity from year to year.)

— MILAN KUNDERA

Philosophers ponder it and pornographers proffer it. Asked why people desire physical beauty, Aristotle said, "No one that is not blind could ask that question." Beauty ensnares hearts, captures minds, and stirs up emotional wildfires. From Plato to pinups, images of human beauty have catered to a limitless desire to see and imagine an ideal human form.

But we live in the age of ugly beauty, when beauty is morally suspect and ugliness has a gritty allure. Beauty is equal parts flesh and imagination: we imbue it with our dreams, saturate it with our longings. But to spin this another way, reverence for beauty is just an escape from reality, it is the perpetual adolescent in us refusing to accept a flawed world. We wave it away with a cliché, "Beauty is in the eye of the beholder," meaning that beauty is whatever pleases us (with the subtext that it is inexplicable). But defined this way, beauty is meaningless—as Gertrude Stein once said about her childhood home, Oakland, California, "There is no there there."

In 1991, Naomi Wolf set aside centuries of speculation when she said that beauty as an objective and universal entity does not exist. "Beauty is a currency system like the gold standard. Like any economy, it is determined by politics, and in the modern age in the West it is the last, best belief system that keeps male dominance intact." According to Wolf, the images we see around us are based on a myth. Their beauty is like the tales of Aphrodite, the judgment of Paris, and the apple of discord: made up. Beauty is a convenient fiction used by multibillion-dollar industries that create images of beauty and peddle them as opium for the female masses. Beauty ushers women to a place where men want them, out of the power structure. Capitalism and the patriarchy define beauty for cultural consumption, and plaster images of beauty everywhere to stir up envy and desire. The covetousness they inspire serves their twin goals of making money and preserving the status quo.

Many intellectuals would have us believe that beauty is inconsequential. Since it explains nothing, solves nothing, and teaches us nothing, it should not have a place in intellectual discourse. And we are supposed to breathe a collective sigh of relief. After all, the concept of beauty has become an embarrassment.

But there is something wrong with this picture. Outside the realm of ideas, beauty rules. Nobody has stopped looking at it, and no one has stopped enjoying the sight. Turning a cold eye to beauty is as easy as quelling physical desire or responding with indifference to a baby's cry. We can say that beauty is dead, but all that does is widen the chasm between the real world and our understanding of it.

Before beauty sinks any deeper, let me reel it in for closer examination. Suggesting that men on Madison Avenue have Svengali-like powers to dictate women's behavior and preferences, and can define their sense of beauty, is tantamount to saying that women are not only powerless but mindless. On the contrary, isn't it possible that women cultivate beauty and use the beauty industry to optimize the power beauty brings? Isn't the problem that women often lack the opportunity to cultivate their other assets, not that they can cultivate beauty?

As we will see, Madison Avenue cleverly exploits universal preferences but it does not create them, any more than Walt Disney created our fondness for creatures with big eyes and little limbs, or Coca-Cola or McDonald's created our cravings for sweet or fatty foods. Advertisers and businessmen help to define what adornments we wear and find beautiful, but I will show that this belongs to our sense of fashion, which is not the same thing as our sense of beauty. Fashion is what Charles Baudelaire described as "the amusing, enticing, appetizing icing on the divine cake," not the cake itself.

The media channel desire and narrow the bandwidth of our preferences. A crowd-pleasing image becomes a mold, and a beauty is followed by her imitator, and then by the imitator of her imitator. Marilyn Monroe was such a crowd pleaser that she's been imitated by everyone from Jayne Mansfield to Madonna. Racism and class snobbery are reflected in images of beauty, although beauty itself is indifferent to race and thrives on diversity. As Darwin wrote, "If everyone were cast in the same mold, there would be no such thing as beauty."

Part of the backlash against beauty grew out of concern that the pursuit of beauty had reached epic proportions, and that this is a sign of a diseased culture. When we examine the historical and anthropological literature we will discover that, throughout human history, people have scarred, painted, pierced, padded, stiffened, plucked, and buffed their bodies in the name of beauty. When Darwin traveled on the *Beagle* in the nineteenth century, he found a universal "passion for ornament," often involving sacrifice and suffering that was "wonderfully great."

We allow that violence is done to the body among "primitive" cultures or that it was done by ancient societies, but we have yet to realize that beauty brings out the primitive in every person. During 1996 a reported 696,904 Americans underwent voluntary aesthetic surgery that involved tearing or burning their skin, shucking their fat, or implanting foreign materials. Before the FDA limited silicone gel implants in 1992, four hundred women were getting them every day. Breast implants were once the province of porn stars; they are now the norm for Hollywood actresses, and no longer a rarity for the housewife.

These drastic procedures are done not to correct deformities but to improve aesthetic details. Kathy Davis, a professor at the University of Utrecht, watched as more than fifty people tried to persuade surgeons in the Netherlands to alter their appearance. Except for a man with a "cauliflower nose," she was unable to anticipate which feature they wanted to alter just by looking at them. She wrote, "I found myself astounded that anyone could be willing to undergo such drastic measures for what seemed to me such a minor imperfection." But there is no such thing as a minor imperfection when it comes to the face or body. Every person knows the topography of her face and the landscape of her body as intimately as a mapmaker. To the outside world we vary in small ways from our best hours to our worst. In our mind's eye, however, we undergo a kaleidoscope of changes, and a bad hair day, a blemish, or an added pound undermines our confidence in ways that equally minor fluctuations in our moods, our strength, or our mental agility usually do not.

People do extreme things in the name of beauty. They invest so much of their resources in beauty and risk so much for it, one would think that lives depended on it. In Brazil there are more Avon ladies than members of the army. In the United States more money is spent on beauty than on education or social services. Tons of makeup— 1,484 tubes of lipstick and 2,055 jars of skin care products—are sold every minute. During famines, Kalahari bushmen in Africa still use animal fats to moisturize their skin, and in 1715 riots broke out in France when the use of flour on the hair of aristocrats led to a food shortage. The hoarding of flour for beauty purposes was only quelled by the French Revolution.

Either the world is engaged in mass insanity or there is method in this madness. Deep inside we all know something: no one can withstand appearances. We can create a big bonfire with every issue of *Vogue*, *GQ*, and *Details*, every image of Kate Moss, Naomi Campbell, and Cindy Crawford, and still, images of youthful perfect bodies would take shape in our heads and create a desire to have them. No one is immune. When Eleanor Roosevelt was asked if she had any regrets, her response was a poignant one: she wished she had been prettier. It is a sobering statement from one of the most revered and beloved of women, one who surely led a life with many satisfactions. She is not uttering just a woman's lament. In *Childhood, Boyhood, Youth*, Leo Tolstoy wrote, "I was frequently subject to moments of despair. I imagined that there was no happiness on earth for a man with such a wide nose, such thick lips, and such tiny gray eyes as mine. . . . Nothing has such a striking impact on a man's development as his appearance, and not so much his actual appearance as a conviction that it is either attractive or unattractive."

Appearance is the most public part of the self. It is our sacrament, the visible self that the world assumes to be a mirror of the invisible, inner self. This assumption may not be fair, and not how the best of all moral worlds would conduct itself. But that does not make it any less true. Beauty has consequences that we cannot erase by denial. Beauty will continue to operate—outside jurisdiction, in the lawless world of human attraction. Academics may ban it from intelligent discourse and snobs may sniff that beauty is trivial and shallow but in the real world the beauty myth quickly collides with reality.

This book is an inquiry into what we find beautiful and why—what in our nature makes us susceptible to beauty, what qualities in people evoke this response, and why sensitivity to beauty is ubiquitous in human nature. I will argue that our passionate pursuit of beauty reflects the workings of a basic instinct. As George Santayana has said, "Had our perceptions no connection with our pleasures, we should soon close our eyes to this world . . . that we are endowed with the sense of beauty is a pure gain." My argument will be guided by cutting-edge research in cognitive science and evolutionary psychology. An evolutionary viewpoint cannot explain everything about beauty, but I hope to show you that it can help explain a good many things, and offer a perspective on the place of beauty in human life.

What Is Beauty and How Do We Know It?

We are always sizing up other people's looks: our beauty detectors never close up shop and call it a day. We notice the attractiveness of each face we see as automatically as we register whether or not they look familiar. Beauty detectors scan the environment like radar: we can see a face for a fraction of a second (150 msec. in one psychology experiment) and rate its beauty, even give it the same rating we would give it on longer inspection. Long after we forget many important details about a person, our initial response stays in our memory.

Beauty is a basic pleasure. Try to imagine that you have become immune to beauty. Chances are, you would consider yourself un-well—sunk in a physical, spiritual, or emotional malaise. The ab-sence of response to physical beauty is one sign of profound depres-sion—so prevalent that the standard screening measures for depression include a question about changes in the perception of one's own physical attractiveness.

But what is beauty? As you will see, no definition can capture it entirely. I started by mining what those who peddle beauty as a business had to say, thinking they might have concrete details about their criteria rather than airy abstractions to float. Aaron Spelling, creator of "Baywatch" and "Melrose Place," said, "I can't define it, but I know it when it walks into the room." I talked with a modeling agency that books top male models, and they were more descriptive: "It's when someone walks in the door and you almost can't breathe. It doesn't happen often. You can feel it rather than see it. I mean someone you literally can't walk past in the street." It is noteworthy that the experts describe the experience of seeing beauty, and not what beauty looks like. On that end, all I got was that they should be young and tall and have good skin. But it was a start.

The Oxford English Dictionary defines the word "beautiful" as "Excelling in grace of form, charm of coloring, and other qualities, which delight the eye and call forth admiration: a. of the human face and figure: b. of other objects." As a secondary definition it states, "In modern colloquial use the word is often applied to anything that a person likes very much." The dictionary that my computer net-work provides says that beauty "gives pleasure to the senses or plea-surably exalts the mind or spirit."

The dictionaries define beauty as something intrinsic to the ob-ject (its color, form, and other qualities) or simply as the pleasure an object evokes in the beholder (The philosopher Santayana called beauty "pleasure objectified.") If we follow a time line of ideas on beauty, the pendulum clearly swings from one direction to the other. For the ancient Greeks, beauty was like a sixth sense. In the twentieth century, when Marcel Duchamp could make a toilet the subject of high art, and Andy Warhol could do the same for a soup can, beauty came to reside not in objects themselves but in the eye that viewed those objects and conferred beauty on them.

Although the *object* of beauty is debated, the experience of beauty is not. Beauty can stir up a snarl of emotions but pleasure must always be one (tortured longings and envy are not incompatible with pleasure). Our body responds to it viscerally and our names for beauty are synonymous with physical cataclysms and bodily obliteration—breathtaking, femme fatale, knockout, drop-dead gorgeous, bombshell, stunner, and ravishing. We experience beauty not as rational contemplation but as a response to physical urgency.

In 1688, Jean de La Bruyère expressed these transgender wishes, "to be a girl and a beautiful girl from the age of thirteen to the age of twenty-two and then after that to be a man." There is tremendous power in a young woman's beauty. In 1957, Brigitte Bardot was twenty-three years old and had starred in the film *And God Created Woman*. That year, the magazine *Cinémonde* reported that a million lines had been devoted to her in French dailies, and two million in the weeklies, and that this torrent of words was accompanied by 29,345 images of her. *Cinémonde* even reported that she was the subject of forty-seven percent of French conversation! In 1994, the model Claudia Schiffer spent four minutes modeling a black velvet dress on Rome's Spanish Steps. According to British journalists covering the "event" for the *Daily Telegraph*, four and a half million people watched and the city came to "a standstill."

Perhaps these are media-driven frenzies, no more real than the canned laughter chortling from our television screens. But small epiphanies are common in daily life. The most lyrical description of an encounter with beauty—solitary, spontaneous, with an unknown other—comes in James Joyce's *Portrait of the Artist as a Young Man* when Stephen Dedalus sees a young woman standing by the shore with "long, slender bare legs," and a face "touched with the wonder of mortal beauty." Her beauty is transformative and gives form to his sensual and spiritual longings. "Her image had passed into his soul for ever and no word had broken the holy silence of his ecstasy. . . . A wild angel had appeared to him, the angel of mortal youth and beauty, an envoy from the fair courts of life, to throw open before him in an instant of ecstasy the gates of all the ways of error and glory. On and on and on and on!"

Ezra Pound had a moment of recognition that inspired him to write a two-line poem "In a station at the Métro," which comprised these brief sentences: "The apparition of these faces in the crowd: Petals, on a wet, black bough." Later, Pound described how he came to write it. "Three years ago in Paris I got out of a Métro train at La Concorde, and saw suddenly a beautiful face, and then another and another, and then a beautiful child's face, and then another beautiful woman, and I tried all day to find words for what this had meant to me, and I could not find any words that seemed to me worthy or as lovely as that sudden emotion. . . . In a poem of this sort one is trying to record the precise instant when a thing outward and objective transforms itself or darts into a thing inward and subjective."

It is difficult to put into words why a particular set of eyes or a certain mouth move us while others do not. Even for the poets, it is often beyond language. Looking to the object of beauty, we confront centuries of struggle to capture beauty's essence.

An Ideal of Beauty Exists in the Mind, Not the Flesh

People judge appearances as though somewhere in their minds an ideal beauty of the human form exists, a form they would recognize if they saw it, though they do not expect they ever will. It exists in the imagination. Emily Dickinson, spending most of her time in her parents' attic, once wrote about the power of the imagination to envision the beautiful: "I never saw a moor, I never saw the sea, Yet I know how the heather looks, And what a wave must be." Kenneth Clark observed in *The Nude* that every time we criticize a human figure, for example that the neck is too short, or the nose too long, or the feet too big, we are revealing that we hold an ideal of physical beauty. Albrecht Dürer wrote that "there lives on earth no one beautiful person who could not be more beautiful."

Donald Symons, an anthropologist at the University of California at Santa Barbara, related this Cartesian experience to me. He attended a talk given by a plastic surgeon in southern California. The surgeon accompanied his talk with a series of slides of very beautiful people. What impressed Symons was that each of these individuals was very beautiful but imperfect. He couldn't help but notice an upper lip that was too long or a nose that seemed too sharply angled. In fact, he felt that their beauty threw this "flaw" into bold relief. But, he wondered, too long or too angled compared to what? For Symons, the experience of looking at such strikingly beautiful faces and seeing these minor deviations from "perfection" was compelling evidence that we possess an innate beauty template which we are unlikely to access directly but against which we measure all that we see. These faces almost matched it, but not quite. Like Dürer, he could envision them being more beautiful.

The human image has been subjected to all manner of manipulation in an attempt to create an ideal that does not seem to have a human incarnation. When Zeuxis painted Helen of Troy he gathered five of the most beautiful living women and represented features of each in the hope of capturing and depicting her beauty. There are no actual descriptions of Helen, nor of other legendary beauties such as Dante's Beatrice. Their faces are blank slates, Rorschach inkblot tests of our imaginings of the features of perfect beauty.

In cinema and in magazines, modern Zeuxises create images of beauty out of the ideal parts of many. Hollywood uses body doubles for stunts requiring a grace and athleticism that actors may not possess. But just as often they do it because someone else's great body looks better matched to that actor's or actress's great face. Jennifer

Beals rose to fame in the 1980s film, *Flashdance*, although it was later revealed that the closeups of her body were not of her body. And it seemed not to matter in the long run. Most people easily melded the face of Beals with the body of her double and kept this composite image in their imaginations.

Top models are genetic freaks whose facial and bodily proportions are well designed to excite and please. But even they bear the marks of human imperfection. Supermodel Cindy Crawford's wrists are different sizes (not to mention her mole!) and supermodel Linda Evangelista hates her mouth because it is "tiny" and "frowny." But there are individuals who have "perfect" feet or hands or lips, and these "specialty models" work full time at modeling only the perfect part. Their hands are placed with the faces of models such as Cheryl Tiegs and Lauren Hutton. The hand market is further specialized into "glamour" hands and "product" hands. The glamour ones have to have great skin and long tapered fingers—"the sort of hands made to wear jewelry and use that American Express card." Product hands are action hands, handling detergents or shampoo bottles with dexterity and steady nerves. Feet are another area of specialty modeling, especially because top models are on average between five feet nine and five feet ten inches tall—they have big feet. For centuries, the ideal foot has been small and delicate, the foot of Cinderella. Foot models have size six (American) feet, with smooth skin and perfect little toes that look like "five little shrimp," as one agency explained.

Of course people come as indivisible packages and the alternate approach to combining the perfect parts of many is to primp and pose one individual into the most pleasing vision possible. Kenneth Clark has written that the naked body is difficult to make into art by a direct rendering. A human body is "not like the tiger or the snow landscape . . . naked figures do not move us to empathy but to disillusion and dismay. We do not wish to imitate, we wish to perfect." This was the approach to portraiture until modernism changed the way bodies were represented. In its most extreme form, images were so idealized that they bore only a cursory resemblance to their subjects. Portraits of the sixteenth-century Queen Elizabeth I rendered her face as "an opaque and unblemished mask." When Horace Walpole was asked to identify true portraits of her, his criteria were the presence of a roman nose, hair laced with jewels, a crown, and a splendid costume of rich fabric, an enormous ruff, and "bushels" of pearls. Elizabeth's portraits were probably never lifelike, but as she aged they became increasingly abstract, focusing attention on the beauty of her spectacular clothing and sketching her face in a shorthand of red-gold hair, pale skin, and a nose with a prominent bridge.

Watch a person looking in the mirror and you will see a person trying to please himself. If we pose for ourselves, we surely always pose for others, attempting to display ourselves as we want to be seen. Icons of beauty just take this several steps further. They undergo elaborate treatment before each appearance, each photograph. In the 1930s screen actresses were presented in dramatic makeup, fabulous clothes, and striking poses in front of filtered lenses. The artifice was obvious and the glamour up front. Today we think we favor natural beauty but natural beauty is as much an artifice as glamour. As model Veronica Webb said when asked how long it took to make her natural beauty, "Two hours and two hundred dollars. . . . I could never make myself look the way I do in a magazine."

In a world where we provide fake, vivid color, airbrushing, and now digital alterations to pictures of everything imaginable, it hardly seems surprising that we want to doctor images of people. We attempt to make everything look better so as to please and to tempt. And we would be fools not to want to please and tempt one another.

Modern artists present us with images stripped of glamour. Diane Arbus photographed people not considered beautiful in unblinking closeups. Photographer Richard Avedon shot a famous series of portraits from the American West, all starkly realized. Painters such as Lucien Freud and Phillip Pearlstein show human flesh with wrinkles, freckles, pallor, and body flab exposed. But these may not be more accurate representations of people—of how our eyes see them, or how they see themselves. We don't usually view people under photographer's lights or get close enough to see all their pores and stray hairs. There is no reason to think that these images are any more "real" than more flattering images. They are cast in the cold light of a surgeon's operating theater, seen through the eyes of the voyeur or your worst enemy. When we look at people we love, or even like, do we ever see them exactly like this? It is just art imposing a different artifice, pretending that we ever view others as just piles of mortal flesh.

Paul Valéry would say that we suffer from the "three-body problem," and can never resolve it. One body is the one we "possess," that we live in. It is for each of us, he says, "the most important object in the world." This is the self that we experience. The second body is the public facade, "the body which has form and is apprehended in the arts, the body on which materials, ornaments, armor sit, which love sees or wants to see, and yearns to touch." We can call the second body the subject of traditional artistic portrayal. The third body is the physical machine that we know about "only for having dissected and dismembered it . . . nothing leads us to suspect a liver or brain or kidney." It is the body we are most estranged from and that beauty covers and helps us to deny.

The reason we have a universal passion for adornment, the reason that photos are doctored and painted representations idealized, is that we long to be not only works of nature but works of art. We want to unite Valéry's three bodies into a unified whole. In part, the longing is spiritual: to have an outer representation that matches our dreams and visions and moral aspirations. It is also a quest for love and acceptance, to have a face and a body that other people want to look at and know. Biologists would argue that at root the quest for beauty is driven by the genes pressing to be passed on and making their current habitat as inviting for visitors as possible. Quentin Bell writes in his stunning book, *On Human Finery*, that painters and dressmakers are all philosophers at heart. "Aristotle said that drama was more philosophical than history for history tells us only what did happen whereas drama tells us what ought to have happened. In this sense the dressmaker and the painter are philosophers. The painter seeks to recreate the body in a state of perfection; the dressmaker seeks to arrange drapery so beautifully that the actual body becomes a mere starting point."

The Beauty Canon

Running as a common thread through the discourses on beauty, from pre-Socratic times onward, is an aesthetic based on proportion and number. The irreducible elements are clarity, symmetry, harmony, and vivid color. Plato said that beauty resided in proper measure and proper size, of parts that fit harmoniously into a seamless whole. He extended the idea of proportion to the beautiful in all things and wrote of the best length of a speech, the optimal organization of paintings, and the proper use of language in poetry. To St. Augustine, beauty was synonymous with geometric form and balance. He thought that equilateral triangles were more beautiful than scalene triangles because their parts were more even. Squares, being composed of equal-length segments, were more beautiful still, circles even more beautiful, and the point, indivisible and pure, was the most beautiful of all. "What is beauty of the body?" he asked. "A harmony of its parts with a certain pleasing color." For Aristotle, beauty resided in "order and symmetry and definiteness." For Cicero, it was "a certain symmetrical shape of the limbs combined with a certain charm of coloring." For Plotinus, it was a "symmetry of parts toward each other and towards a whole . . . the beautiful thing is essentially symmetrical." Plotinus believed that beauty must be present in details as well as the whole; "it cannot be constructed out of ugliness, its law must run throughout." Common to all these theories is the idea that the properties of beauty are the same whether we are seeing a beautiful woman, a flower, a landscape, or a circle.

Artists throughout history have tried to capture the geometric proportions of beauty by devising measurement systems for the human body. As art historian George Hersey has noted, the most important human proportion system in Western art dates to the fifth-century Greek Polyclitus, whose sculptures of a male spear bearer and wounded female Amazon represent much-imitated standards for representation of the human male and female form. Polyclitus's contemporary Praxiteles articulated a similar female paragon in his Aphrodite of Cnidos. These bodily canons influenced all of Western art from approximately 450 B.C. to the early twentieth century, until modernism expanded our representations of the body. Polyclitus called his male spear bearer the Canon, and so it has remained.

For Polyclitus, and later for Albrecht Dürer, Leon Battista Alberti, and Leonardo da Vinci, beauty resided in symmetry. For these artists and theoreticians, symmetry had a different meaning than it does today. When we speak of symmetry we mean exact correspondence of form on opposite sides of a dividing line or plane or central axis. To the Greeks and the Renaissance artists and scholars, symmetry meant the relation between, and the exact correspondence among parts, usually expressed in whole or rational numbers. It meant, as George Hersey has pointed out, "commensurability." So, for example, the whole body was measured in hand heights or head lengths or in relation to thumb length. Galen argued that an arm corresponding to three hand lengths is more symmetrical and hence more beautiful than one corresponding to two and a half or three and a half hand lengths.

Dürer used his own finger as the unit of measure to construct a proportional system in which the length of the middle finger equaled the width of the hand, and the width of the hand was proportional to the forearm. From there, he constructed a canon for the whole body. His entire system for measuring ideal beauty rested on the proportions of his hands, which were very long-fingered. We might wonder what would have happened to Western art if Dürer had had small fingers! But this is not an isolated example of an artist or scientist incorporating his own features into a universal canon. Edward Angle published a classic set of orthodontic indices in 1907 in which he used his own (European) face as the ideal. This meant that all Asians and Africans would have needed to have their teeth straightened!

During the Renaissance, particular attention was paid to the proportions of the ideal human face as well as those of the body. Dürer proposed that the face in profile separates into four equal divisions, while others proposed a division into thirds with equal space from the hairline to the eyebrow, from the brow to the lower edge of the nostrils, and from the nostrils to the chin. Other neoclassical and Renaissance guidelines dictated that the height of the ear and the nose be equal, that the distance between the eyes equal the width of the nose, that the width of the mouth be one and a half times as wide as the nose, and that the inclination of the nose bridge parallel the axis of the ear. These rules dictated the representation of beauty in Western art for centuries, and in the twentieth century are the highly influential bases from which plastic surgeons pillage to resculpt and reconstruct faces.

The canons are revered in Western civilization. But surprisingly few people have been interested in scientifically testing whether they describe the actual proportions of living beauties. However, anthropometrist Leslie Farkas took out his calipers and measured the facial proportions of two hundred women, including fifty models, as well as young men and children, and had large numbers of people rate their beauty. Then he compared his measurements and the beauty ratings with the ideals of the classical canon. His results are hardly definitive, but they provide some intriguing information. The canon did not fare well. Many of the measures did not turn out to be important, such as the relative angles of the ear and nose. Some seemed pure idealizations: none of the faces and heads in profile corresponded to equal halves or thirds or fourths. Some were inaccurate—the distance between the eyes of the beauties was greater than that suggested by the canon (the width of the nose). Farkas's results do not mean that a beautiful face will never match the Renaissance and classical ideals. But they do suggest that classical artists might have been wrong about the fundamental nature of human beauty. Perhaps they thought there was a mathematical ideal because this fit in a general way with platonic or religious ideas about the origin of the world.

Measurement systems have failed to turn up a beauty formula. Perhaps it isn't surprising that universal beauty does not conform to the ratio of Dürer's finger. In fact, as we will see, beauty may come from a mathematically messy set of criteria having more to do with our biology than with ideal numbers.

Beauty Satanic and Divine

No single attitude about beauty has been consistent throughout history. People have revered beauty, they have scorned it and loathed it. Plato believed that beauty made the spiritual visible. Sensual beauty imitates pure beauty, which we cannot access. Beauty, like truth and justice, is a platonic Pure Form, of which things of this world may offer us glimpses but never truly incarnate. This is how Plato explained beauty's strange power, its mysterious ability to awaken aesthetic bliss. As Thomas Mann wrote in *Death in Venice*, all virtues would inspire reverence if we could but see them: "beauty alone is . . . the only form of the spiritual which we can receive through the senses. Else what would become of us if the divine, if reason and virtue and truth, should appear to us through the senses? Should we not perish and be consumed with love, as Semele once was with Zeus?"

With the arrival of Christianity, the attitude toward beauty became more ambivalent. Church leaders grappled with the right way to respond to it. "There is nothing good in the flesh," St. Clement said, "the man of god must mortify the works of the flesh." Jerome saw the flesh as something to be "conquered." The teaching of Christ told his followers to renounce temptation and the transient things of this world. Beauty was feared as a sensual temptation and a worldly vanity. But it was also revered as an image of God's grace. According to Genesis, man is made in the image of God, therefore his appearance is divine, and the more beautiful, the more Godlike. "Beauty is the mark of the well made, whether it be a universe or an object," said Thomas Aquinas, and the well made is an "imitation of an idea in the mind of the creator." The history of Judeo-Christian attitudes toward beauty reflects an agonized struggle to reconcile beauty as temptation and beauty as God's glory. In Dürer's four books of human proportion, released after his death in 1528, he speaks of the physical perfection of Apollo, Adam before the Fall, and Christ. Their perfect beauty is a sign of their divinity, while our imperfect beauty is a sign of our fall from grace.

Attitudes toward beauty are entwined with our deepest conflicts surrounding flesh and spirit. We view the body as a temple, a prison, a dwelling for the immortal soul, a tormentor, a garden of earthly delights, a biological envelope, a machine, a home. We cannot talk about our response to our body's beauty without understanding all that we project onto our flesh.

Psychoanalysis assumed a legacy of shame around the body. Freud wrote, "The love of beauty seems a perfect example of an impulse inhibited in its aim." That is, beauty derives from sexual excitement that must be deflected away from its source. "It is worth remarking that the genitals themselves, the sight of which is always exciting, are nevertheless hardly ever judged to be beautiful." Too much cultivation of beauty, he wrote, reflects pathological narcissism. Like masochism and passivity, narcissism is largely a female problem, a cover for shame and worthlessness, feelings to which women are prone.

Until recently, many people who sought cosmetic surgery ended up getting psychiatric diagnoses—they were labeled depressed, hysterical, obsessional, narcissistic. If the patient was a man, he was almost always given a psychiatric diagnosis since attention to appearance in a man was considered a graver sign than it was in a woman. In the last twenty years, the number of "healthy" recipients of plastic surgery has vastly increased according to psychiatric studies. Perhaps this is a reflection of more mainstream acceptance of plastic surgery and a greater diversity among its clients. But it is equally likely to reflect a change in modern psychiatry, which can look at appearance enhancement as something other than an unhealthy need. Psychoanalyst John Gedo recently made the radical suggestion that cosmetic surgery is not so different from altering character traits by means of psychoanalysis: both are attempts at refashioning the self. Psychiatrist Peter Kramer has made analogies between cosmetic surgery and what he calls "cosmetic psychopharmacology," for example the use of drugs such as Prozac not just to cure depression, but to transform personality, to feel "better than well."

The Evolution of Beauty

The social sciences have been strangely absent from the rich intellectual debate about the nature of human beauty. As you will see, much of the research that informs the arguments of this book emerged only in the 1970s and after. Gardner Lindzey's 1954 *Handbook of Social Psychology*, a lengthy tome devoted to the study of social interaction, listed only one entry for "physical factors." Any reading of psychology and anthropology texts written before the late 1960s would suggest that physical appearance had absolutely no bearing on human attitudes or affections, and no role in human mental life. Why have the social sciences had so little interest in the human body?

One reason is that the social sciences were not interested in the biological "givens." As anthropologist John Tooby and psychologist Leda Cosmides have pointed out, the standard social science model (SSSM) that developed over the past century viewed the mind as a blank slate whose contents were determined by the environment and the social world. The mind itself was believed to consist of a few general-purpose mechanisms for perceiving and understanding the environment. It was a model that divided biology from culture, and then ignored biology (the mere slate) to probe the influential work of culture. The roots of the model within the social sciences of this century are political and social as well as intellectual.

As anthropologist Donald Symons says, you cannot understand what a person is saying unless you understand whom they are arguing with. Cultural relativism came to the intellectual forefront particularly in the United States during the 1920s as a reaction to claims that races, ethnic groups, classes, women, and so on were innately inferior. Such arguments were countered with evidence from behaviorism, showing that people can drastically alter their behavior in response to environmental rewards and punishments. As John B. Watson, the founder of behaviorism, wrote, "Give me a dozen healthy infants, well-formed, and my own specified world to bring them up in and I'll guarantee to take any one at random and train him to become any type of specialist I might select—doctor, lawyer, artist, merchant-chief, and yes even beggar-man and thief, regardless of his talents, penchants, tendencies, abilities, vocations, and race of his ancestors."

Similarly, the SSSM presented evidence from other cultures to show that human behavior was malleable, plastic, and largely or wholly acquired through experience. Margaret Mead's idyllic description of the sexual freedom of Samoan girls was in this tradition. In this context, it is not surprising that the most entrenched belief about beauty among social scientists was that "beauty is in the eye of the beholder." Focusing on the range and inventiveness of human adornment, from brass rings that create giraffe necks, to painted teeth and lip plates, they concluded that beauty must be a matter of individual taste or cultural dictate.

Gardner Lindzey brings up another reason that beauty was shunned by social scientists—the "spectacular failure" of previous attempts to link physical attributes to behavior (phrenology, physiognomy, and so on). In the next chapter we will review these studies and see that they yielded very little in the way of scientific fact and spread many fictions. It is no wonder that many scientists were eager to dissociate themselves from this work. Charles Darwin was one of many of its near victims. The captain of the *Beagle*, like many people of his time, had been influenced by the physiognomist Johann Caspar Lavater's *Essays on Physiognomy*, written in 1772, which suggested that certain facial features predict character. As Darwin wrote in his biography, the captain "was an ardent disciple of Lavater . . . and he doubted whether anyone with my nose could possess sufficient energy and determination for the voyage." As psychologist Leslie Zebrowitz has said, "The theory of evolution was almost lost for want of a proper nose."

Social scientists shunned beauty as trivial, undemocratic, and all in all not a proper subject for science. But by the late 1960s, Lindzey was chiding his colleagues for their "neglect of morphology [outward appearance]" and suggesting, "Perhaps now is the time to restore beauty and other morphological variables to the study of social phenomena." Within the next three decades an explosion of research was to provide compelling evidence for a new view of human beauty. It suggested that the assumption that beauty is an arbitrary cultural convention may simply not be true.

The research comes at a time when scientists have begun to question anew many other assumptions about the relationship between human behavior and culture. As Leda Cosmides, John Tooby, and Jerome Barkow point out: "Culture is not causeless and disembodied. It is generated in rich and intricate ways by information-processing mechanisms situated in human minds. These mechanisms are in turn the elaborately sculpted product of the evolutionary process." Clearly, culture cannot just spring forth from nowhere; it must be shaped by, and be responsive to, basic human instincts and innate preferences. Until the 1960s it was believed that languages could vary arbitrarily and without limit, but now there is a consensus among linguists that there is a universal grammar underlying this diversity. Similarly, it was thought that facial expressions of emotion could arbitrarily vary across cultures until the psychologist Paul

Ekman showed that many emotions are expressed by the same facial movements across cultures. Ekman made the important distinction between the facial expression of emotion (smiles, frowns, scowls, and so on), which are universal, and the rules for when to display those emotions, which show cultural variation. Similarly, aspects of judgments of human beauty may be influenced by culture and individual history, but the general geometric features of a face that give rise to perception of beauty may be universal.

Of course, no one is suggesting that people are conscious of the evolutionary rationale behind their aesthetic reactions, just that these are the pressures that shaped those reactions as the human brain evolved. Nor is anyone suggesting that learning and culture do not play any role in our judgments of beauty. As poet Charles Baudelaire wrote in the nineteenth century, beauty is made up of an "eternal invariable element" and a "relative, circumstantial element," the latter defined as "the age, its fashions, its morals, its emotions." "I defy anyone," he wrote, "to point to a single scrap of beauty which does not contain these two elements."

Putting beauty into the realm of biology completely alters the time frame of our analysis. Recent feminist writings on beauty, such as Naomi Wolf's *The Beauty Myth*, have been criticized by Camille Paglia and others for being ignorant of history, because they look at images of beauty only in this century, not throughout the thousands of years of human civilization. Paglia herself claims that beauty was invented in ancient Egypt. The premise of this book is that beauty's history is far, far longer! The ability to perceive beauty and respond to it has been with us for as long as we have been men and women.

As Cosmides and Tooby have said, "The time it takes to build circuits that are suited to a given environment is so slow it is hard to even imagine—it's like a stone being sculpted by wind-blown sand. Even relatively simple changes can take tens of thousands of years." Our minds are products of a long history and a vanished way of life. For ninety-nine percent of the history of our species we lived as hunter gatherers in nomadic bands of small numbers. To understand our instincts, we must turn backward in time and place our minds *in habitat.*

In the following pages we will look at the argument for beauty as a biological adaptation. The argument is a simple one: that beauty is a universal part of human experience, and that it provokes pleasure, rivets attention, and impels actions that help ensure the survival of our genes. Our extreme sensitivity to beauty is hard-wired, that is, governed by circuits in the brain shaped by natural selection. We love to look at smooth skin, thick shiny hair, curved waists, and symmetrical bodies because in the course of evolution the people who noticed these signals and desired their possessors had more reproductive success. We are their descendants.

Of course, such signals are now manipulated by cosmetics, plastic surgery, and clothing, three giant industries in part devoted to false advertising. Additionally, one cannot escape a comment on the irony of sexual attraction: in a world where men and women try to stave off pregnancy for the majority of their sexual encounters, sexual preference is still guided by ancient rules that make us most attracted to bodies that look the most reproductively fit. Nor can we escape the jarring thought that women compete in the mating world for men whose brains are hard-wired to find nubile teenagers highly desirable and particularly beautiful. This is not a conscious process nor a desired one but a biological holdover from a vanished way of life. Is it resistible? The reaction to beauty may be automatic, but our thoughts and our behaviors are ultimately under our control.

We begin to look at the science of beauty in Chapter 2 by focusing on the least controversial aspect of the theory: why we find babies irresistibly attractive. We will also look at more controversial research suggesting that parents respond more affectionately to physically attractive newborns. Finally, we will review the research on infants' perceptions and see that even at three months of age they are gazing longer at attractive faces than at unattractive faces. Infants appear to come into the world equipped with the ability to discriminate and prefer the beautiful. This has been some of the most powerful research showing that beauty preferences are not learned.

In the next two chapters we will examine the powerful impact of beauty in everyday life. Beauty influences our perceptions, attitudes, and behavior toward others. Economist David Marks has said that beauty is as potent a social force as race or sex. But unlike racism and sexism, which we are conscious of, "lookism," or beauty prejudice, operates at a largely unconscious level. These studies put some of our extreme beauty practices in perspective. People are spending billions of dollars on cosmetics and plastic surgery for a reason: these industries cater to a world where looking good has survival value.

Although most people would say they no longer believe that "what is beautiful is good," preferential treatment of beautiful people is extremely easy to demonstrate, as is discrimination against the unattractive. From infancy to adulthood, beautiful people are treated preferentially and viewed more positively. This is true for men as well as women. Beautiful people find sexual partners more easily; and beautiful individuals are more likely to find leniency in the court and elicit cooperation from strangers. Beauty conveys modest but real social and economic advantages and, equally important, ugliness leads to major social disadvantages and discrimination. Do beautiful people end up being happier? The answer may surprise you.

In Chapters 5, 6, and 7 we look at beauty itself. Much of the world's beauty from the peacock's tail to the nightingale's song functions as a courtship signal calling attention to the bearer's physical splendor. Humans are no exception and we find beautiful those physical characteristics that suggest nubility, fecundity, health, and good design. What are those signals? Anthropologists and psychologists have suggested that a beautiful face is—average. That is, beautiful faces display the features of the population mean. Further research has suggested that, while average is attractive, the most beautiful faces are not average but have features that deviate from the average in a small number of predictable ways. These advocates of the "more is better" school suggest that, like the flamboyant peacock's tail, we advertise our fitness by exaggeration.

Research on the body has also provided nonintuitive and surprising results. For example, symmetry, proportion, and in particular the ratio of waist to hip size, can be more powerful determiners of the beauty of a woman's (and a man's) body than absolute weight (barring extremes of obesity or emaciation). I will touch briefly on the controversy surrounding the role of the media in eating disorders. With more than one third of the United States population obese, and the number growing steadily, there is no indication that the plethora of thin beauties is creating a society of thin people at all. The media contribute to greater dissatisfaction with real bodies by crowding our minds with examples of extreme body types. But eating disorders have a more complex origin.

Some skeptics may argue that studies of the psychophysics of beauty reinforce the idea that there is "one ideal of beauty." This is a misunderstanding. In fact, the theory predicts only that certain geometric proportions of the face and body and a few exaggerated features are beautiful. It also predicts that certain features will be universally seen as unattractive. The model can be incarnated in a dizzying variety of physical types rather than cookie cutter copies.

Next, we will take a look at fashion shifts and crazes. Despite their instability and our constant cravings, fashions may change but they never get more beautiful. In fact, the most coveted fashions can end up looking ridiculous and find their way into secondhand bins quickly. We will look at the driving engines of fashion—sex and status—as well as the ways that fashion reflects aesthetic and personal as well as social aspirations.

We will end by trying to put beauty in perspective, first by looking at other potent forms of human nonverbal communication, including smell, and then by considering beauty in the larger circumference of human life.

Feminists, and thoughtful women and men in general, are deeply ambivalent about beauty. It is seen both as a source of strength and as a source of weakness and enslavement, something that blinds others to our deeper nature. The desire of women in particular to have their totality recognized is deeply felt. Beauty may be a "pure gain," but its social effects, from harassment of the beautiful to discrimination against the unbeautiful, to the neglect of less visible "inner" beauty, may be anything but positive.

How to live with beauty and bring it back into the realm of pleasure is a task for twenty-first-century civilization. The difficulties women face are a sign of the inherent misfit between the conditions in the ancient world in which we evolved and current civilization. But the solution cannot be to give up a realm of pleasure and power that has been with us since the beginning of time.

CHAPTER TWO

Beauty as Bait

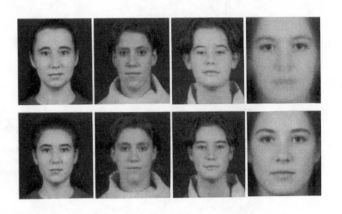

Beauty is a greater recommendation than any letter of
introduction.

—ARISTOTLE

'I cannot say often enough how much I value beauty as a quality
that gives power and advantage. . . . It takes first place in human
relations; appears in the foreground, seduces and prepossesses our
judgments, exercises great authority, and is marvellously
impressive.

—MICHEL DE MONTAIGNE

Unfortunately I could not help listening to him; he was handsome
as the dawn.

—CATHERINE THE GREAT

Many people have an idyllic conception of childhood as a time when beauty does not matter. Listen to children taunt and tease each other in a schoolyard—shrimp, squirt, four eyes, fatso—to quickly disabuse yourself of that notion. Children gravitate to beauty. One of photographer Richard Avedon's first snapshots was of his seven-year-old sister Louise. The nine-year-old Avedon was so entranced by her that he taped the negative to his skin and had the sun burn it into his shoulder. Her oval face, dark hair, big eyes, and long throat became "the prototype of what I considered to be beautiful. She was the original Avedon beauty." His later photographs of models Dovima, Suzy Parker, Dorian Leigh, and Carmen Dell'Orefice "are all memories of Louise."

Children are sensitive to beauty from a very early age, but how and when do they acquire their preferences? The popular wisdom is that children learn beauty preferences through acculturation. Perhaps their parents foist certain tastes upon them, then peers rebelliously revise the aesthetics, and pop culture finally fine-tunes it. As Robin Lakoff and Raquel Scherr wrote in their 1984 book *Face Value*, "Beauty is not instantly and instinctively recognizable: we must be trained from childhood to make those discriminations."

But psychologist Judith Langlois is convinced that no lessons are required: we are born with preferences and even a baby knows beauty when she sees it. Langlois collected hundreds of slides of people's faces and asked adults to rate them for attractiveness. When she presented these faces to three- and six-month-old babies, they stared significantly longer at the faces that adults found attractive. The babies gauged beauty in diverse faces: they looked longer at the most attractive men, women, babies, African-Americans, Asian-Americans, and Caucasians. This suggests not only that babies have beauty detectors but that human faces may share universal features of beauty across their varied features.

Langlois is quick to point out that infants show preferences for beautiful unfamiliar faces. It is unlikely that an infant's behavior toward his or her caregivers is influenced by their facial beauty, given the importance of attachment to the baby's survival. Nor is she suggesting that babies with attractive mothers have a special eye for beauty. Babies looked longer at attractive faces regardless of the mother's attractiveness.

The notion that infants come prewired with beauty detectors was not the prevailing theory when Judith Langlois began her research ten years ago. The idea that an infant would be peering out at the world with the eyes of a neonate beauty judge is downright discomfiting: even *they* notice looks? But her results are part of a growing body of evidence that infants share a universal set of sensual preferences. They prefer to look more at symmetrical patterns than at asymmetrical ones, and to touch soft surfaces rather than rough ones. By four months of age they prefer consonant to dissonant music. When psychologists Jerome Kagan and Marcel Zentner played dissonant melodies to babies, they wrinkled their noses in disgust. Kagan and Zentner felt that they were witnessing the first signs of a preference for easy listening and mellifluous crooning. We can learn to love dissonance, but it is an acquired taste.

Babies pay close attention to the human face. Within ten minutes of emerging from the mother's body, their eyes follow a line drawing of a face. By day two they can discriminate their mother's face from a face they have never seen before. The next day they begin mimicking facial actions: stick out your tongue at a newborn and the baby will do the same. Each newborn orients immediately toward whatever is biologically significant, and topmost will be people who ensure her survival.

Babies look almost as long at a person's eyes as they do at the whole face, and see there much of what they need to know. The movements of the eyes and of the muscles surrounding the eyes, the changes in pupil size, and the gleam or dullness in our eyes express nuances of feeling. The small individual differences in distances around the eyes created by the facial bone structure is one of the most enduring parts of our visual signature, and as unique as fingerprints. Automatic face recognition systems guided by computers recognize faces better from the eyes alone than from the nose or mouth alone. Computers learning to detect faces from nonfaces are most easily fooled by interference with the eye regions. This is why masking only the area around the eyes has proved an effective disguise from Don Juan in the fourteenth century to the Lone Ranger in the twentieth.

If babies see someone looking at them, they look back, and usually they smile. Their interest piqued, they will look up to three times as long at a face looking at them as at a face looking away. Unlike prey animals such as rabbits and deer which have panoramic, surround vision, humans, like hawks and leopards and other predators, look precisely at what they are thinking about. This is why babies come equipped with mechanisms to detect direction of gaze, and why the human eye may have evolved its distinctive appearance. Unlike most animals, which have sclera that darken with age, humans retain white sclera all of their lives. The whites of the eyes help us gauge where eyes are looking and give us a good idea of what has captured other people's attention and what might be on their minds.

An animal stalked by lions, which can see prey from a mile away, would not be greatly benefited by seeing the whites of their eyes. By then, it's all over. But for humans living in close proximity and dependent on one another for survival, direction of gaze is an effective form of communication, whether in the form of the predatory gaze, the beseeching look, or the look of love.

The newborn baby's preferences are *formes frustes* of adult preferences. Babies turn into adults who like symmetry and harmony and things that feel smooth; they are riveted by the sight of the human face, and aroused when eyes meet theirs. The three-month-old who stares at beautiful faces grows up to be the usual person whose head is turned by the sight of beauty and who can fall in love by looking. When babies fix their stare at the same faces adults describe as highly attractive, their actions wordlessly argue against the belief that culture must teach us to recognize human beauty.

Cuteness

Meanwhile, the baby is being sized up by adults. Fifty years ago the ethologist Konrad Lorenz suggested that infant features set off a rush of tender emotions. Infants come equipped with many triggers. They have soft skin and hair, huge eyes, big pupils, chubby cheeks, and small noses. Their heads are big and their limbs are small and springy. Infants would die without our care, and so they had better arrive irresistible.

The reaction to baby features is automatic and we behave tenderly toward any creature who mimics them. Toy manufacturers and cartoonists capitalize on our innate preferences for juvenile features. Born in the 1930s, Mickey Mouse began his life so debonair and lithe that writer Graham Greene described Fred Astaire in the movie *Top Hat* as "the human Mickey." But over the next fifty years Mickey aged in reverse. His eyes and head kept getting bigger while his limbs kept getting shorter and thicker. Mickey now mimics the basic geometry of the human infant. Disney's Bambi has the exaggerated high forehead of an infant, as well as those doe eyes.

Cute features are accidents of biological development. The baby's brain and nervous system develop early, and the baby's eyes are almost adult size at birth while the hands and feet are tiny miniatures of what they will become. Although there is nothing intrinsically beautiful about huge eyes and small hands, the basic baby construction has profound meaning to us and elicits tender feelings. Just as chicks have stripes, lion cubs have spots and tail rings, and baby chimps have white tail tufts, humans have big heads and big eyes, chubby cheeks and little noses that signal their helplessness. Jane Goodall found that baby chimps were safe from attack as long as they retained the tuft of white on the tail, suggesting that the white tail tufts are biological tags warning adults not to harm the little chimp. The visible labels that humans wear in their first year of life may serve the same purpose: to turn off aggression.

Queen Victoria, mother of nine, once said, "An ugly baby is a very nasty object." Perhaps she was just voicing her Victorian disdain for untidy, ill-mannered creatures. For most people, there are no ugly babies, just as there are no ugly puppies or ugly brides. All babies are cute, or at least cute to their parents, who find them so from the second they are born. As Anna Quindlen said, "Who was it coined that old saw about God making them so cute so we will not kill them? It has particular resonance at 4 A.M."

However, it turns out that there are slight differences in the way mothers act during their newborns' first few days of life, and some of this behavior is in response to the baby's appearance. Psychologists videotaped mothers and infants within days after the baby's birth and then three months later. They also had a separate set of observers look at color photographs of the babies and rate their attractiveness.

They found that the mothers of the most attractive newborns spent the most time holding the baby close, staring into the baby's eyes and vocalizing to the baby. They needed to be forcibly unglued in order to pay attention to anyone else. The mothers of the less attractive newborns spent more time tending to the baby's needs (wiping, burping, checking, adjusting, and so on) and had their attention deflected more easily. They were not neglectful, but they seemed more reserved in their affection and a little less swept off their feet.

The researchers did not ask the mothers to rate their infants' looks, perhaps because they thought it too incendiary. But they did ask the mothers a number of questions about the baby and about child care. The mothers of the less attractive babies were more likely to report feelings of stress, complain about lack of time and energy, and worry about money. These differences largely disappeared by three months, although mothers were still being more affectionate and playful with their attractive baby boys at this age.

People find it easy to make judgments about which babies are the cutest (this is why there are infant beauty contests and Gerber babies). Beautiful babies are typical babies, or babies whose features mildly exaggerate the typical baby geometry: they just push all the triggers. Babies considered ugly do not have all the triggers, and it makes them look older, as if some version of their future face had been placed on their newborn body. Babies born prematurely, like many at-risk babies, have these falsely mature faces. When their pictures are mixed in with pictures of full-term babies, these babies are imagined to be difficult and irritable, and people are less willing to volunteer to take care of them. In fact, babies who are not cute can suffer even more dire consequences. When abused children under court protection were studied in California and Massachusetts, it turned out that a disproportionate number of them were unattractive. This wasn't because they were badly groomed or bore unhappier facial expressions than other children. Rather, abused kids had head and face proportions that made them look less infantile and cute. Such children may be more likely to suffer abuse because their faces do not elicit the automatic reaction of protection and care that more infantile faces do. These children may also be perceived as more capable than they are and may be more subject to unrealistic expectations because of their older appearance. There is evidence that abusing parents often do have unrealistic expectations of their children. Finally, their unusual appearance may signal poorer health or viability, as it may with a premature infant.

In the animal world, newborns who advertise their health and viability get more attention from their mothers. American coots are grayish-black birds whose chicks have vivid orange plumes and bald heads that turn bright red during feeding. Chicks beg for food visually, flashing their red and orange signs for their mother. If researchers cut back the orange plumes, the drabber chicks get less attention and less food from the mothers, who feed the colorful, good-looking chicks first. The mothers appear to be ignoring the ones assumed to be not healthy enough to display their colors.

Such behavior is not worlds away from mothers of high-risk, low-birth-weight twins in a United States suburb who were studied by psychologist Janet Mann. Mann observed that by eight months the mothers showed clear preferences for one of the twins, spending more time soothing, holding, playing, and vocalizing with her. It did not seem to matter which infant cooed, smiled, or followed the mother more, all mothers favored the healthier twin. Although mothers usually provided roughly equal care and feeding to both babies, in two cases where the families were extremely poor, the sicker twin appeared to be severely neglected. Mann concluded that the mothers had developed preferences (that were not conscious) for the twin most likely to survive, preferences driven by mechanisms which over the course of evolution have maximized a mother's reproductive fitness.

The difficult truth is that throughout evolutionary history parents have faced limited and uncertain resources, and at-risk babies have been less likely to survive. A mother sensitive to cues of her baby's health and viability would be more reproductively successful knowing how much to invest in this new baby without endangering herself and the lives of her other children. This is not a cold calculation but a realistic consequence of tragic circumstance.

Parents today may never face such heartbreaking dilemmas. They may have the resources and the safety to invest time and energy in a child at serious risk. But, as we can see from close observations of mothers with newborns, they must still override evolved mechanisms from an ancient brain to do so. Parents still tend to favor healthy babies and respond most affectionately to babies who look most like typical, big-eyed, small-nosed, round-cheeked bundles. In the ancestral environment, the baby's appearance was the best early diagnostic indicator of whether the baby would survive, and whether or not to unleash to-die-for love.

Father Knows Best

Parents may be cautious before giving away their hearts, but once they've sensed that their baby is viable, they open all the doors. Remember that the infant's looks are most important in the first days of life. By then, if not before, parents believe that, among many other sterling attributes, their baby is better-looking than anyone else's. Parents and family members also peer into a baby's face to see whom he or she resembles. Immediately after the baby's birth, mothers are apt to say that the baby resembles the father.

Psychologists Margo Wilson and Martin Daly sent questionnaires to hundreds of new mothers and fathers and their relatives. They found that claims of paternal resemblance were very common, and were significantly more common than claims of maternal resemblance. Indeed, in many families "everyone" commented on the baby's resemblance to the father.

Daly and Wilson interpret their findings this way: mothers have no doubt that the baby is theirs, but fathers always run some risk of being duped. Before DNA testing, fathers had two sources of information: their knowledge of the mother's fidelity, and physical evidence from the appearance of the baby. Facial features are highly heritable. Emphasizing the baby's resemblance to the father helps to erase any doubts and stoke his affection and investment in the newborn. Mothers respond to healthy cuteness; fathers also want to know, Does she look like me? Seeing some reflection of his own features in the baby's face is a powerful trigger of paternal feelings.

In the 1920s anthropologist Bronislaw Malinowski went to the Trobriand Islands in the South Pacific. Trobrianders believed that the mother is impregnated by spirits, not semen. Yet people suggested that children resembled the "father" (the mother's husband) more than their mothers or siblings. It was even considered bad manners to suggest that a child resembled the mother.

Daly and Wilson suggest that fathers everywhere are sensitive to whether or not the baby resembles them and mothers everywhere encourage the father's belief in the baby's resemblance. Such behaviors probably operate automatically. They appear even if the man doesn't have a clue as to how babies are conceived, as is the case with the Trobrianders. Daly and Wilson warn that there may be a dark side to the father's desire for a baby who looks like him. For example, in families in which one child is singled out for abuse, it may well be the child least resembling the father. Adoptions are more successful when parents perceive the child to be similar to themselves. Daly and Wilson predict that this factor will be more important to adoptive fathers than to adoptive mothers.

Of course, baby looks are general crowd pleasers. Women's pupils and children's pupils dilate involuntarily when they see pictures of babies. No one begrudges babies their beauty nor would most doubt that their beauty is an adaptive solution to an evolutionary problem: how to guarantee their survival. Babies teach us that responses to physical beauty are automatic, and irresistible, that they start early and run deep.

Appearance and Reality

Life would be much easier if we retained the compelling power of appearance that we have as babies. As we leave infancy, we lose the protection cuteness affords. Our white tail tuft gone, we face the world unshielded: adult beauty is a great advantage, but it protects the few, not the many.

We face a world where lookism is one of the most pervasive but denied of prejudices. People like to believe that looks don't matter. But every marketing executive knows that packaging and image are as important as the product, if not more so. We treat appearance not just as a source of pleasure or shame but as a source of information.

Minds are not designed to disentangle surface and substance easily: deep down, few people believe that the relation between the two is accidental or arbitrary. Young children find it especially difficult to separate appearance from reality. When psychologists show a young child a squirrel, and then shave it and paint it so that it looks like a raccoon, the child will say that it is now a raccoon. They are so swayed by appearance that they forget that the squirrel is still there beneath the shaved and painted exterior.

There is a good evolutionary reason why we place so much value on appearance. Looks have been a reasonable and sometimes solitary guide to what is good and what is bad for us. Brown spots and wrinkled skin tell us that fruit is past nutritional peak, and green color may tell us that it is unripe. Biologist George Orians believes that we share universal preferences for particular landscapes because they signal safety and refuge. In collaboration with Judith Heerwagen, he surveyed painters, gardeners, photographers, and others about what kinds of landscapes they consider beautiful, Orians and Heerwagen found that all were attracted to landscapes with large trees, views of the horizon, water, changes in elevation, and multiple paths out. Geographer Jay Appleton suggests that such environments offer prospect and refuge, the ability to monitor the outside world from a safe place.

What do we look for in other people? For centuries, people thought that the human face forecast character and personality. As Tolstoy lamented, "It is amazing how complete is the delusion that beauty is goodness." As we will see, goodness, at least moral goodness, has nothing to do with it.

Looking Good

The idea that carnal beauty is visible evidence of spiritual beauty can be traced back at least as far as Plato, who believed that mortal beauty was a reflection of ideal beauty. Sappho wrote that "what is beautiful is good." These ideas flourished again among the Renaissance humanists. Marsilio Ficino saw beauty as "the blossom so to speak of goodness. By the allurements of this blossom, as though by a kind of bait, the latent interior goodness attracts all who see it. . . . We would never know the goodness hidden away in the inner nature of things, nor desire it, unless we were led to it by manifestations in exterior appearance." Baldassare Castiglione wrote in 1561 that "beauty is a sacred thing . . . only rarely does an evil soul dwell in a beautiful body, and so outward beauty is a true sign of inner goodness . . . it can be said that in some manner the good and the beautiful are identical, especially in the human body. And the proximate cause of physical beauty is, in my opinion, the beauty of the soul." Sociologist Anthony Synott remarks that Castiglione's ideas represent "a superb synthesis of biology and theology, the profane and the sacred, sex and God." They revel in the beauty of the physical body and call it worship of the soul.

Ugliness was a sign of the bad, mad, or dangerous. Deformities, ugliness, and disease were seen as stigmas branded onto the body by a wrathful God. Castiglione said, "For the most part the ugly are also evil." In the sixteenth century Francis Bacon wrote, "Deformed persons are . . . (as the Scripture saith) void of natural affection."

Attempts to specify the particular facial characteristics that give clues to character date back to Aristotle. In 1586, Giovanni Della Porta, an Italian naturalist and philosopher, wrote a treatise called *De humana physiognoma* in which he attempted to understand the relationship between body and soul in humans. Since human personalities are opaque and complex, he found it useful to draw analogies to animals that have a simpler psychology. His scheme was based on the magical belief that things that look alike are alike. Each animal has a defining passion and, by analogy, each human who resembles an animal shares this passion. An ass is foolish, a mule is stubborn, a rabbit is timid, an ox is dumb, and a pig is dirty and greedy. Della Porta argued that if a man bore resemblance to a particular animal "let him be aware that he will behave in a similar fashion." In other words, if you look like an ass you will act like an ass, and if you look like a fox, a fox you are.

In writings from Plato onward, the straight profile of the Greek statue was usually assumed to be the ideal human face. One of its many assets was that it did not resemble the faces of rabbits, goats, apes, frogs, or any other ignoble animals. If beauty meant not looking like a beast, the Apollo Belvedere was the gold standard. Discovered in Rome around 1496, the ancient Greek statue of Apollo (dated to approximately 320 B.C.) was what art historians call "the totemic statue of High Renaissance art in Italy." (The Belvedere got its name from the villa behind the old Vatican palace where the papal families of the sixteenth and seventeenth centuries housed their collection of the great marble statues of Greece and Rome.)

For the eighteenth-century philosopher Hegel, the Greek profile "far from being an outer and accidental form is the incarnation of the very idea of beauty itself. . . . Thanks to it we have an embodiment of a facial conformation in which the expression of the spiritual takes pride of place." Hegel based this on the fact that Greeks had a prominent nose bridge, which in profile traced a continuous line from the thinking centers (the forehead) to the face, and thereby focused visual attention on the upper rather than the lower, sensual parts of the face.

Eighteenth-century Dutch artist and anatomist Petrus Camper invented a device to measure facial angles from profiles. He measured the face horizontally from the ear to the lip, and vertically from the most protruding point of the forehead to the most forward point of the upper jaw (usually at the upper lip). The intersection of these points was the facial angle. Camper's facial angle became the first widely used measurement system for comparing the skulls of different races. But Camper's intention was to try to quantify beauty. He took for granted that the statues of Greek antiquity represented the ideal of beauty. As he wrote, "We will not find a single person who does not regard the head of Apollo or Venus as possessing a superior beauty, and who does not view these heads as infinitely superior to those of the most beautiful men and women." Camper found that the ancient Greek statues had facial angles of 100 degrees (relatively straight profiles) while most human profiles range from 70 to 90 degrees. Since monkeys, dogs, and other animals have facial angles lower than humans and Greek statues had angles slightly higher, Camper thought he had found the angle of beauty. As he wrote, "What constitutes a beautiful face? I answer a disposition of traits such that the facial line makes an angle of 100 degrees with the horizontal." Measuring the skulls of different races, he found that facial angles increased from orangutans and monkeys to African blacks to orientals to the European man and finally to the Greek statue, making European man the closest to the beauty ideal and African man the furthest. A Swiss pastor, Johann Caspar Lavater, published his own set of facial angles, again in ascending order, this time from the frog to the Apollo Belvedere, again placing European man closest to the ideal of beauty.

By such analogies and comparisons, European men put forth the idea that European men and women were the most beautiful of humans. Since their facial angles approximated those of the Greek gods, so did their character and intelligence. The work had a tribal imperative and was used to justify cultural or racial superiority. Of course, the profile of the Apollo Belvedere is not more beautiful than the profile of a handsome African man, appearance is not reality, analogies do not prove anything, and racism is a clear example of a skin-deep fallacy. Scientists now agree that the concept of race explains very little about human variation. Beneath the skin, less than seven percent of genetic differences can be explained by belonging to one race or another. There is more genetic diversity within races than between them. Dissimilar appearances can cloak sisters under the skin, and similar appearances may cover widely divergent personalities. Racism is real, race may not be.

These days, we know that it is difficult to get a read on moral character, intelligence, or soul just by looking. If Mother Teresas always looked like Miss Universe, the world would be just and appearance would be an easy read. But no one has figured out the visible signs of saint or sinner. It is sometimes convenient to liken a human feature to an easily visualized animal one, such as doe eyes, but we use it as a visual not moral descriptor. Knowledge about personality and character is something that we gather slowly from what a person does to us and to others. We say now that physical beauty is but "skin deep" and that "pretty is as pretty does."

But we don't always act that way. What pretty does is often seen in a forgiving light. We are quick to leap to judgments of the unbeautiful, imagining for instance that fat people are lazy or greedy. We know that the link between beauty and goodness is spurious, but our actions are not always guided by conscious reason.

The Injustice of the Given

Whether or not the beautiful is good, beauty seems to bring out goodness in others. In one psychologist's study, seventy-five college men were shown photographs of women, some of whom were very attractive and others less so. They were asked to select the person they would be most likely do the following for: help move furniture, loan money, donate blood, donate a kidney, swim one mile to rescue her, save her from a burning building, and even jump on a terrorist hand grenade. The men were most likely to volunteer for any of these altruistic and risky acts for a beautiful woman. The only thing they seemed reluctant to do for her was loan her money.

Answers to psychologists' questions about hypothetical situations may have little to do with real behavior. But when put to the test, at least in small ways, people seem to confirm what the college boys say. In several staged experiments, psychologists have tested people's honesty and altruism toward good-looking and plain-looking people and find that their good deeds are not doled out evenly. For example, in one study a pretty or an ugly woman approaches a phone booth and asks the occupant, "Did I leave my dime there?" (There is a dime in the phone booth.) Eighty-seven percent of people return the dime to the good-looking woman, but only sixty-four percent return the dime to the ugly woman. In another study, two women stand by a car with a flat tire in the roadway: the good-looking one gets rescued first.

People are more likely to help attractive people even if they don't like them. In another staged experiment, an attractive or unattractive woman gave men compliments on their work or criticized it. Afterward, the men were asked how much they liked the woman. They particularly liked the attractive woman who praised them, and liked least the attractive woman who criticized them. But asked to volunteer more time, the men gave it to the good-looking woman, even when he didn't like her. As the psychologists wrote, her attractiveness attracted. Attractiveness attracts even in situations where there is no chance of actually meeting the recipient of one's favors. In yet another study, completed (bogus) college applications were left in Detroit airports. A note attached to them suggested that the applications were given to fathers who had accidentally left them behind. Each had the identical application answers, but each had a different photograph attached. People were much more likely to mail the applications of the better-looking applicants.

Interestingly, people are *less* likely to ask good-looking people for help. This is particularly true for men with good-looking women, but it is also true for both men and women with good-looking members of their own sex (it is less true for women asking good-looking men for help). But as evolutionary psychologists Leda Cosmides and John Tooby have shown, people keep a watchful eye on who has done what for whom. Our efforts to please good-looking people with no expectation of immediate reward or reciprocal gesture are one way we reinforce beauty as a form of status, not unlike being born into the nobility or inheriting wealth. Beauty represents what writer Jim Harrison has called "the injustice of the given."

The high status of beauty is one reason why it is a subject fraught with such heated emotions. Didn't democratic societies ban the aristocracy and level the playing field? Perhaps this is also why we are so easily persuaded by the idea that beauty is attainable through the usual democratic means—hard work and money. If it confers elite status, then we must make it an elitism based on effort and achievement, not a priori advantage. Historian Lois Banner has chronicled "the democratic rhetoric of beauty experts in the early twentieth century," which insisted that "every woman could be beautiful." She suggests that such campaigns were dangerous for women because they held up an unattainable ideal. Estee Lauder's successful campaigns included her exhortations that "there are no homely women only careless women . . . you have to want it [beauty] very much and then help it along with some well-chosen products." Paradoxically, the arguments of twentieth-century beauty experts have often unwittingly linked beauty with goodness. Women who were dissatisfied with what they saw in the mirror now felt not only unattractive but lazy, inept, or lacking the inner beauty which was supposed to shine forth with good habits and good concealer.

Beauty as Status

As we walk down the street, we negotiate space with other people. We carry a small territory with us, a protected turf that surrounds us whether we are sitting or standing, and upon which others cannot trespass without permission. Move in too close, and people get uncomfortable. Tall people have bigger territories: their sheer size intimidates people. When people are asked to approach a stranger and stop when they no longer feel comfortable, they will stop about two feet away from a tall person (22.7 inches to be exact) but less than a foot (9.8 inches) from a short person. Very attractive people of any size are given big personal territories; they carry their privileges around their persons.

Good-looking people are more likely to win arguments and persuade others of their opinions. People divulge secrets to them and disclose personal information. Basically, people want to please the good-looking, making conciliatory gestures, letting themselves be persuaded, telling them informative gossip, and backing off from them, literally, as they walk down the street.

But perhaps people are awed by their confidence and assertiveness, not their looks. Perhaps they persuade by intelligence or force of personality. In fact, attractive people do tend to be more at ease socially, more confident, and less likely to fear negative opinions than are unattractive people. They are more likely to think that they are in control of their lives rather than pawns of fate and circumstance, and they are apt to be more assertive. In a particularly interesting study, people were asked to participate in an interview with a psychologist. During the course of the interview the psychologist was interrupted by a colleague and excused herself. If the interviewee waited patiently, the interruption would last ten minutes. Attractive people waited three minutes and twenty seconds on average before demanding attention. Less attractive people waited an average of nine minutes. There was no difference in how the two groups rated their own assertiveness. Attractive people merely felt entitled to better treatment.

We know that behavior often becomes a self-fulfilling prophecy. After all, it would not be surprising if beauties, who are deferred to, agreed with, and granted favors, automatically assumed the privileges of power. Even ten minutes of being treated the way beautiful people are treated can bring changes in behavior. Psychologists set up a study in which women and men talked on the telephone for ten minutes; during this time the men were told to try to get to know the women. Each man had been given a Polaroid snapshot of the woman they were supposed to be talking with. In their mind's eye, the phone companion was beautiful or ugly. In fact, all men were talking to the same woman. The really interesting part of this study is that the woman became more animated and confident in conversations with men who believed her to be good-looking. The men were trying harder, and were bolder, sexier, and funnier, and they brought out the bold and sexy in her. She sounded attractive when she was presumed to be attractive.

Status is a prized commodity that we confer on the beautiful. As we will see, we believe that the beautiful have many things that we want, and that they may be in a position to help us get them too.

To Whom Much Is Given, Much Is Expected

We expect attractive people to be better at everything from piloting a plane to being good in bed. We guess that their marriages are happier, their jobs are better, and that they are mentally healthy and stable. For practically any positive quality you can think of, people will assume that good-looking people have more of it, do it better, and enjoy it more.

The expectations start in childhood. Teachers in four hundred classrooms in Missouri were given a report card of a fifth-grade student, including grades, evaluation of attitude, work habits, and attendance. The only variant was the attached photograph of the child—an attractive or unattractive boy or girl. Despite the wealth of information about behavior and performance, looks swayed opinions. The teachers expected the good-looking children to be more intelligent and more sociable and popular with their peers. What is more disturbing is that good-looking students often do get better grades. When the subjective aspects of grading are removed, and grades are based solely on standardized tests, the advantage disappears.

Interestingly, despite the "dumb blonde" stereotype, people presume that attractive people of both sexes are more, not less, intelligent than unattractive people. This is particularly true for males, but the effect holds for females as well. It helps to explain the finding of several different studies that the better looking the fill-in-the-blank (painter, essayist, student) is, the more positively his or her work is evaluated. In these studies, looks help most when the work is not of high quality, but they can even give an extra boost to good work. Social psychologists call this a halo effect, after the luminous aura that surrounds revered figures.

And truth be told, we still have a hard time thinking that something evil dwells in these temples. Karen Dion, one of the pioneer researchers on the effects of attractiveness, asked adults to consider hypothetical seven-year-olds who step on dogs' tails or throw icy snowballs at other children's heads. When a good-looking child is depicted doing these things, adults give the child the benefit of the doubt and presume that the child is having a bad day or is the victim of circumstance. The adults don't believe that he or she has done this before or will do it again. The unattractive children are more likely to be eyed suspiciously as possible future juvenile delinquents.

Good-looking adults are more likely to get away with anything from shoplifting to cheating on exams to committing serious crimes. They are less likely to get reported (they aren't being eyed suspiciously), and if they are reported, they are less likely to get accused or penalized. Law enforcement officials, juries, and judges don't just judge the current circumstances and the person's past behavior but they take a look and think: Could she have done that? Just as teachers can take the same report cards and come to different predictions based on appearance, so can judges and juries. This effect is particularly strong for attractive females.

Occasionally good looks can backfire, and when they do it tells us something about the expectations we have for good-looking people. Swindling is the crime of defrauding a person of money or property. The stereotyped swindler is a smooth talker (usually male) offering get-rich-quick schemes, or the femme fatale of many a film noir whose mercenary motives are opaque (or irrelevant) to her besotted target. If accused of this crime, the good-looking fare worse than the less attractive. They look good enough to pull it off, and they are punished for abusing the power of their beauty.

Beauty is an advantage in all realms of life. But it is important to realize the magnitude of the advantage. In most studies, attractive people have an edge but it is small to moderate rather than large. Most studies compare great-looking people with very unattractive people, whereas people tend to hover near the average. The studies say as much about the disadvantages of being below average in looks as about the advantages of beauty. In fact, the evidence is that the penalty for ugliness might be even greater than the reward for beauty.

Packing Heat

But in the sexual domain the importance of looks cannot be over-estimated. People expect attractive people to be very popular, so-cially confident, and at ease. They also expect them to be sexually exciting, responsive, experienced, and adventurous. Men expect beautiful women to have a high sex drive and prefer variety in sex. People assume that good lookers of both sexes have more dates, fall in love more often, and start their sex lives earlier.

As we've seen, the expectations for social confidence and social ease are true, even if they are just self-fulfilling prophecies. Even four-year-olds and ten-year-olds desire good-looking children as friends. Once they reach dating age, both good-looking men and women are more popular with the opposite sex. They have more dates, more opportunities for dalliance, and they get more attention. Friendship is another story. Good-looking women in particular en-counter trouble with other women. They are less liked by other women, even other good-looking women.

Imagine that you are talking to an attractive stranger when a much more attractive stranger enters the room. Chances are, the person you are talking to suddenly looks a shade less attractive. Psy-chologists call this a "contrast effect," and men seem more suscepti-ble to it than women are. After staring at photographs of very attrac-tive faces, men show less desire to date an average-looking woman. Exposure to extremely beautiful bodies in visual erotica can wreak havoc on some men's judgment. In one study, men shown pictures of women's beautiful bodies in erotica rated a previously attractive nude as less exciting. Some men even claimed to be less in love with their wives! Although it is unlikely that love can be shaken by a glance at a photograph, the men's responses suggest the temporary power of an image of a stunning body. We have a chance to calibrate faces against real-life faces all the time. Everyone sees hundreds and even thousands of faces, but most people have not seen hundreds of nude bodies. Given a much smaller database, nude or minimally clothed bodies in the media may get disproportionate representations in our minds and skew our ideas of the possible or even of the average.

This may be one reason why women often dislike it when their mates indulge in watching pornography. It most certainly is one reason why beautiful women have a tougher time holding on to female friends. We try to control our social environment to make ourselves look good, or at least better than the other choices, and no one wants her own light dimmed by having a beacon next to her.

In the guppy world, males are the brightly colored and ornamented sex. Male guppies prefer to hang around with other males who will not outshine them. Scientists rigged up a setting in which male guppies could either swim where they wanted (which was next to females) or were kept away from the females by an invisible barrier. Other males watched this and assumed that the males who were swimming far away had been rejected by the female. Later, the observing males were put into the water to roam freely. They spent more of their time around the "rejected" male, presumably hoping to profit by the comparison. They hoped to produce a contrast effect.

What about the sexual mystique of beauty? It turns out that good-looking men and women are more sexually experienced and engage in a greater variety of sexual activities. Both good-looking men and women begin to have sex earlier, although for women this does not necessarily translate into more partners. Studies by scientists Randy Thornhill and Steven Gangestad suggest that good-looking men are more likely to bring their women to orgasm, and to simultaneous orgasm. Intriguing, yes. Sorry to drop it now but we will return to this in Chapter 6 where we discuss the many advantages of the symmetrical body. All of this suggests that the good-looking may indeed be having more fun, at least in bed.

Good-looking people don't have any monopoly on great sexual technique. But they do have more opportunities, and without much effort they've already engaged the fantasies of their partners. As we've seen from the behavior of women who, during ten-minute phone calls, suddenly act more alluring, better performance can easily be coaxed by a partner. It is not uncommon for people to fantasize about sex with a more beautiful stranger while making love to their partners, probably for this very reason.

One of the interesting upshots of work on stereotypes of the attractive is that these stereotypes exist for both sexes. Beauty is an advantage for men as well as for women, although the magnitude is greater for women. We are told that women bear the burden of appearances when, in fact, so do men. But there are some differences that are not small in magnitude. One is that men make many more sexual inferences about women based on appearance than women do. Men are much more likely to believe that attractive women are sexually permissive, high in sex drive, and sexually confident. Women aren't so sure based on appearances alone.

The men's belief may actually be strategic: a sexually permissive woman with a high sex drive might be receptive to advances, and this belief may make it easier for men to approach her. Men have other such strategies. For example, men are much more likely to read friendly gestures as signs of sexual interest and attempts at seduction than women are. Men want to be more promiscuous than women, and often desire more variety in partners. If they believe that they are picking up signals of interest from a woman (whether or not they are) they will be more likely to approach her. Even if this opens up opportunities for sexual contact a small fraction of the time, such beliefs would give the men who hold them a reproductive advantage.

People have high expectations of beautiful adults and children, men and women. But as we see, beauty plays a particularly important role in sex and romance. We will look more closely at this advantage. And after cataloguing the privileges of beauty, it may seem silly to ask if beauty makes us any happier but I'll pose the question anyway.

CHAPTER THREE

Pretty Pleases

Why may I not speak of your Beauty, since without that I could never have loved you.

— JOHN KEATS TO FANNY BROWNE, JULY 8, 1819

If anyone were desirous to produce a being with a great susceptibility to beauty, he could not invent an instrument better designed for that object than sex . . . sex endows the individual with a dumb and powerful instinct, which carries his body and soul continually towards another, makes it one of the dearest employments of his life to select and pursue a companion, and joins to possession the keenest pleasure, to rivalry the fiercest rage, and to solitude an eternal melancholy.

— GEORGE SANTAYANA

To work hard, as I've worked, to accomplish anything, and then have some yo-yo come up and say, "Take off those dark glasses and let's have a look at those blue eyes" is really discouraging.

— PAUL NEWMAN

In the animal world, gaudy plumage and huge body ornaments emerge at sexual maturity, and animals reserve their brightest colors for courting displays. Caterpillars turn into butterflies, and peachicks explode into the psychedelic colors of the peacock when it is time to reproduce. Flowers are alluring landing strips for pollinating insects: they are the plant world's sex objects. Throughout the natural world, beauty is the harbinger of sexual reproduction.

At puberty, male and female bodies shed their string-bean torsos and emerge with new shapes, curves, and angles. Boys' voices deepen, their skin darkens, their muscle mass increases, and their faces acquire dominant jaws and brow ridges. A girl's skin lightens and her lips swell from surges of estrogen. Her body acquires the extra fat it will need to support reproduction, and sequesters it in hips and breasts. Her once tubular torso rounds into an hourglass.

For most of human history, people mated in their teenage years and conceived a first baby by age twenty: being a teenager was all about courting and choosing. In many parts of the world it still is. When anthropologist Suzanne Frayser surveyed four hundred and fifty-four traditional cultures, she found that the highest frequency of brides was in the twelve to fifteen years old age category, and the largest age category for grooms was eighteen.

Girls at this age are preternaturally beautiful. Supermodel Christy Turlington was discovered at age thirteen, Kate Moss when she was fourteen. Back in 1937, Lana Turner became "the sweater girl" when she was fifteen years old. Truman Capote described the eighteen-year-old Holly Golightly as having a face "beyond childhood yet this side of belonging to a woman," on the knife edge between innocence and experience. Adolescence is when we find out how potential mates respond to us. It is not surprising that, as psychologist Mary Pipher writes, "the preoccupation with bodies at this age cannot be overstated," and that adolescents are notoriously hypercritical and extremely attentive to their beauty.

But the preoccupation does not end with adolescence; we continue to be evaluated as mate material all of our lives. When Dustin Hoffman starred in the movie *Tootsie* he became obsessed with making Tootsie not only a credible woman but an attractive one. He was not entirely successful: as Bill Murray said to him in the film, "Don't play hard to get." Hoffman realized "that if I had met me at a party, I would never have gone up to me. I would never have asked me out. And that made me emotional because I thought I'd missed a lot of good—really interesting women in my life. . . ."

In fantasies, everyone possesses an ideal mate. Real behavior is constrained by opportunity, but in our imaginations we can have what we want and do what we want. Fantasy fodder such as *Playboy* and *Playgirl* is filled with images of healthy, young, good-looking people. The audience for this material is over ninety percent male. Romance novels are equally popular, accounting for forty percent of mass market paperback sales, but their narratives of everlasting love are geared to women. One look at the covers suggests that they too are breeding grounds for visual fantasies. Over fifty million romance novel covers display Fabio, a six-foot-three-inch, 220-pound model from Milan with long blond hair, a twelve-month tan, and a broad muscular chest. Fabio first made his appearance in 1986 and has ridden bare chested across covers of romance novels ever since. He now has a promotional calendar the sales of which match those of Cindy Crawford, and a 900 number over which he dispenses romantic advice.

In the real world you don't need to be a Prince of Midnight or a Playmate of the Month to attract a partner, but the initial combustion between potential mates hinges precariously on looks. Social psychologists Elaine Hatfield and Susan Sprecher assert that "at the beginning of a romance there is probably nothing that counts more." Hatfield began her research on attractiveness in the 1960s after organizing an undergraduate dance where she matched half the couples on what she called "social desirability" (intelligence and personality) and the other half randomly. She was clearly missing something since the carefully matched pairs were no happier than the randomly flung-together pairs. Then she took a look at looks. Anyone matched with a good-looking partner was not only happy but took a phone number. Twenty years later, psychologists matched one hundred gay men at a tea dance. They found that virtually all of the men, regardless of their own level of attractiveness, preferred the most attractive partners and wished to see them again.

Preferences based on looks turn up from Australia to Zambia. In 1990 psychologist David Buss interviewed over ten thousand people from thirty-seven cultures between the ages of fourteen and seventy about their mating preferences. Around the world, kindness was a highly valued quality in a mate, but physical attractiveness and good looks were on everyone's top ten list of important and desirable qualities.

Western cultures have been accused of placing extreme value on physical beauty. But in Buss's study, people in more than one third of the non-Western and non-North American countries placed *more importance* on the looks of their mates than did college students in the United States. Buss and his colleague, psychologist Steven Gangestad, found that prevalence of parasitic diseases and not exposure to supermodels was the key factor in determining how much a culture valued physical beauty in a mate. Cultures with a higher prevalence of parasitic diseases are even more likely to place a very high value on physical beauty because beautiful features such as a glorious mane of hair, clear skin, and a lean muscular body are visual certificates of health.

Flamboyant signals of health in populations susceptible to disease are well documented in the animal world. Biologists William Hamilton and Marlene Zuk studied hundreds of songbirds and found that the most brightly colored birds come from the most heavily parasitized populations. If this seems paradoxical, remember that, by the logic of evolution, the species most susceptible to parasites should be the showiest because they are under evolutionary pressure to advertise their hereditary resistance. The bright coloration makes it visually easy to separate the bright and healthy from the dull and parasite-ridden. A heavily parasitized peacock cannot divert the resources necessary to keep up its metabolically expensive tail, any more than a protein- or iron-deficient human can spare the fuel to grow a thick lustrous mane of hair. Extreme displays in disease-ridden populations help to guide mates to the healthiest partners. Hundreds of neotropical birds show the same association between health and bright colors and showy displays.

Most people do not end up with mates who have the glossiest hair or the clearest skin of all potential mates. Instead, couples tend to be very well matched in looks. As the *Sex in America* survey put it, "It's not that you never see a stranger across a crowded room and fall instantly in love. It's more that that stranger you notice will look just like you." In other words, the person will be roughly equivalent to you in beauty. The better one looks, the better-looking one's partner is likely to be; it's another incentive to look as good as possible.

Fair Sex?

More women than men diet and women outnumber men in eating disorders nine to one. Eighty-nine percent of patients for "aesthetic procedures" done by members of the American Society of Plastic and Reconstructive Surgeons in 1996 were women. Women are more likely than men to dye their hair, shop for clothes, wear jewelry and makeup, wear perfume, and pinch their toes into ill-fitting shoes for the sake of beauty.

There are two reasons why this is true: the first is men. Men value the looks of their sexual and romantic partners more than women do—or at least they say and think they do and have said so and thought so for a long time. In 1939 men and women were asked how important good looks are in a marriage partner. On a 0–3 scale, men ranked looks at 1.5, and women ranked them at .94. In 1989, when men and women were asked the same question, looks had become more important for both sexes but the male-female difference remained stable. Men rated the importance of looks as 2.1, and women ranked them at 1.67.

Men value looks more than women do in virtually every culture where the question has been asked. In the 1950s biologists Clelland Ford and Frank Beach found that the physical attractiveness of women received more "explicit consideration" than the physical attractiveness of men in almost two hundred tribal cultures. In his 1990 study David Buss found that men valued physical attractiveness and good looks in a partner more than women did in thirty-four of the thirty-seven cultures he studied. In India, Poland, and Sweden there was no difference between the sexes on the question of good looks. In none of the cultures that he studied did women care more about the looks of their partners than men did.

It is men, not women, who are the major consumers of the eight-billion-dollar pornography industry. According to the *Report of the U.S. Commission on Obscenity and Pornography* the patrons of adult bookstores and movie theaters are "predominantly white, middle-class, middle-age, married males." There is a market for images of naked men too, but it is gay men, not women, who are the consumers. The women's magazine *Viva* presented nude male centerfolds in the 1970s but discontinued them. *Playgirl* magazine conducted a private survey to find out how much women were enjoying their nude male centerfolds. Only a fourth of them enjoyed looking at the centerfolds "a great deal." *Playgirl* is rumored to appeal mainly to a gay male audience, though it denies the rumor.

Pornography has never been able to engage the imaginations of most women the way it does most men. This is because pornography more directly taps male desires—for anonymous sex, for quick sex, for visual thrills. Men are much more likely to fantasize about partners they do not know, and even to substitute different partners during the same fantasy. A man typically has sex with thousands of partners in his imagination over the course of his life. Women are much more likely to fantasize about people they know. All studies of sexual preference show that men find not only the fantasy but the potential reality of sex with a stranger more appealing than women do.

The upshot is that men spend a lot of time staring at women but women do not spend nearly as much time staring at men. In fact, the vast majority of pictures women look at are pictures of attractive women in women's magazines: they are interested in checking out the competition. Donald Symons even suggests that women may like to "girlwatch" naked women in pornography more than they like to "boywatch" naked men. He likens it to learning from "another person's skilled tool-using performances."

People composing personal ads are trying, consciously or unconsciously, to appeal to the mating preferences of the hoped-for partner. Ads placed by "women seeking men" are the most likely to mention their beauty; next most likely are "men seeking men"; and least likely are "women seeking women." Basically anyone seeking a man, be it a gay man or a straight woman, advertises his or her good looks. Anyone seeking a woman is more likely to mention sincerity, friendship, and financial security. As one dating service director said, "Men just look at the pictures, women actually read the things."

Donald Symons suggests that male-male and female-female unions tell us a lot about the evolved psychologies of men and women. They represent male and female sexuality in undiluted form, without the compromises and adjustments that heterosexual partners make to satisfy the other sex. The evidence from same-sex partners suggests that the male interest in a beautiful partner is not just men's way of objectifying and denigrating women. Men interested in men are just as interested in the beauty and youth of their male partners. Women, whether straight or gay, desire beauty but they are less likely to see it as important in a partner.

For both sexes, looks are more important for short-term casual relationships than for more serious relationships. Since men tend to want many more casual partners than most women do, this means that looks are automatically more a part of their psychological landscape. Men's desire for good looks in a short-term partner is off the scale. But women's is not far behind. What about for more serious relationships? On a scale from 3 (indispensable) to 0 (irrelevant or unimportant), men in the United States rate the importance of good looks in a long-term partner at 2.11 and women rate it at 1.67. Across thirty-seven cultures, the sex difference remains—men rate the importance of looks at 1.86 and women rate it at 1.47. This difference does not look big but it is statistically significant and, as we will see, psychologically significant as well. Neither sex sees attractiveness as a neutral or unimportant quality in a mate and neither sees it as indispensable. But both want it, and men want it more.

Young Love

Physical beauty is like athletic skill: it peaks young. Extreme beauty is rare, and almost always found, if at all, in people before they reach age thirty-five. We say that time steals beauty, but time merely moves on from an idealized moment when the body has developed into its physical and (pro)creative powers and has yet to decline in health and fertility. Many tortured quests for beauty are efforts to stretch these few years into perpetuity, to retain the appearance of nubile adolescence forever. As William Butler Yeats wrote, "decrepit age . . . has been tied to me, as to a dog's tail." If only we could capture youthful exuberance and bottle it. We wish to stay forever young not only in our hearts and our minds but in our bodies.

Although "rough winds do shake the darling buds of May" for both sexes, it is women who try more desperately to hold on to youth. This is because they want to appeal to men, who often seek young partners. Gay and straight men rate photographs of younger potential "sex objects" as more attractive than older ones. Straight women tend to prefer slightly older men, and lesbians are neutral on the subject. A two- or three-year age difference between men and women is typical for first marriages, with the bride being younger. For example, the average age of a first-time bride in the United States in 1996 was 24.8 years and for a first-time groom 27.1 years.

Once men get into their thirties they begin to fetishize younger women. If men marry a second time, their wives are on average five years younger than they. If they marry a third time, the new wife is likely to be eight years younger. A man may like a younger woman for many reason including longings for his own lost youth, the desire to play a father figure, and the need to dominate and control, but the mating statistics suggest that marrying younger women may simply reflect men's desire to mate with a maximally fertile woman, or at least one who looks that way.

The media present us with male mating preferences unconstrained by reality. Hollywood movies pair Nick Nolte at age fifty-three with Julia Roberts at age twenty-seven in *I Love Trouble* or Warren Beatty at age sixty-one with Halle Berry at age twenty-nine in *Bulworth.* Hollywood presumes that men are believable romantic leads at any age, and that they are believable when cast as younger men (in other words, that no one is paying very close attention to age cues in male faces or bodies). Although women take far greater pains than men to appear younger, Hollywood's nubility detectors are not fooled. After a woman reaches her mid-thirties, she may be cast as a character older than herself, but she is unlikely to be cast in the role of a younger woman, or even a woman her own age. *The Graduate* paired the stammering and innocent twenty-year-old Benjamin Braddock with the cool, martini-drinking older woman seductress, Mrs. Robinson. Anne Bancroft was thirty-six at the time, Dustin Hoffman was thirty-one. *The Bridges of Madison County* depicted the romance between a middle-aged man and woman. There was considerable opposition to casting the forty-five-year-old Meryl Streep in the role of the forty-five-year-old farmer's wife, and she got the role only after younger actresses were auditioned. However, there was no problem with the male lead, Clint Eastwood, who at sixty-five was actually thirteen years older than the character he portrayed. A man's talent and star power can override concerns about his age, a woman's does not. Hollywood movie makers and their audiences appear to have very specific visual requirements for female romantic leads, and apparent youth is one of them.

Discussing the frequent pairings of aging male stars with young starlets, David Buss was quoted in the *New York Times* as saying, "There is a point where men cease to embody the qualities that women desire, but it's not much shy of a wheelchair and the need for total medical care." Perhaps he had in mind pairings such as the former Playmate and Guess model Anna Nicole Smith to the eighty-nine-year-old wheelchair-bound Texas millionaire she married, J. Howard Marshall II. But partners are usually no more than eight years apart in age. Both sexes are turned on by signs of a healthy partner. Women do not prefer old men, they prefer men who give signs that they have resources and are willing to invest them in her and in potential offspring. Such savoir-faire and power bear an uncertain relation to age. It is men who have a decided age preference: they seek youthful partners.

Marrying Up

Beauty preferences would not be important unless they had consequences in the real world. We can advertise for a dreamboat and pay our money to indulge fantasies, but the key issue is what decisions we make given the choices we have. All the evidence suggests that men's real-life choices are heavily influenced by appearances. The best-looking girls in high school are more than ten times as likely to get married as the least good-looking. Better-looking girls tend to "marry up," that is, marry men with more education and income than they have. A man's looks in high school or his looks at any age do not predict whether he will marry or the financial status of his future mate. In Darwinian terms, the less attractive women get the short end of the reproductive stick. Women who do not marry and women who marry men with fewer resources lower their chances of bearing viable offspring and being able to support them. If the halls of high school haunt women, it's because life does not let them entirely escape the judgments that echoed down them.

There is little evidence that women with greater intelligence have any advantage on the marriage market. On the contrary, in one recent study of more than ten thousand men and women in Wisconsin, women who had never married turned out to be significantly more intelligent than the women who had married.

Being with a good-looking woman ups a man's status. When people are shown pictures of a man with a very attractive woman who is described as his girlfriend, they say that he is more self-confident, intelligent, and likable than when they are shown the same picture but told that the woman is a stranger. As Milan Kundera has said, "Women don't look for handsome men, they look for men with beautiful women." What happens when a woman is with a handsome male mate? Nothing. She is not seen as any smarter or more likable.

The higher status of female beauty in the mating world is reflected in the market for images of objects of desire. Across all professions, women make about seventy cents to a male dollar for the same work. But when it comes to body display, sex discrimination works the opposite way. In 1994, *Forbes* magazine estimated that the top three female models made $6.5 million (Cindy Crawford), $5.3 million (Claudia Schiffer) and $4.8 million (Christy Turlington). Male model salaries usually top in the low six figures: among the highest-paid models, top male models earn ten percent of what females do. For noncelebrity models whose salaries are in the low-five-figures range, female models make about twice what male models do.

Good looks are a woman's most fungible asset, exchangable for social position, money, even love. But dependent on a body that ages, it is an asset that a woman uses or loses. It is as perishable as trigger reflexes, exquisite balance, and quick reaction times. Although beauty is most easily exchanged for male attention—that's obvious and has an obvious explanation, that the men are hoping for a chance to have sex with her—it is interesting that it is also exchanged for interest from women and even children (this is true to a lesser extent for good-looking men as well). This is because beauty is convertible into other assets that people covet, for example, wealth, connections, surplus suitors, and so on.

Beauty Envy

Earlier, I said that there are two reasons women care so much about beauty. The first is men. The second is women. Women's glances can be as scrutinizing as the male gaze, only more critical. The root is female-female competition for males, just as the root of male aggression and competition is male-male competition for females. Science writer Matt Ridley suggests that men and women mold each other, and mold the status hierarchies within each sex: "Pamela Anderson Lee was made by men, just as Mike Tyson was by women." Just as generations of men have wanted voluptuous women and generations of women have wanted alpha males, selection pressures have given us what we have desired.

Women torture themselves about minor beauty flaws, and can't help but compare their looks with those of other women. When the other woman is more beautiful, they feel envious, and may subconsciously try to even the score (she must be dumb or shallow, a bitch or a bore). Talking at MIT, Camille Paglia ridiculed this attitude. "I don't feel less because I'm in the presence of a beautiful person. I don't go, Oh, I'll never be that beautiful! What a ridiculous attitude to take! . . . When men look at sports, when they look at football, they don't go, Oh, I'll never be that fast, I'll never be that strong! When people look at Michelangelo's David, do they commit suicide? No. See what I mean?" It's an interesting point. On the one hand, women admire beautiful women, copy their styles, and allow them top places in the female hierarchy. But they also envy these women, and the envy poisons the pleasure. Envy is hostility toward the very thing one desires.

Why is there so much self-denigration and envy? Because every woman somehow finds herself, without her consent, entered into a beauty contest with every other woman. No matter how irrelevant to her goals, how inappropriate to her talents and endowments, or how ridiculous the comparison, women are always compared one to another and found wanting. Hillary Clinton's haircuts and the length of Marcia Clark's skirts get as much press coverage as their words and actions. Oksana Baiul captured the gold medal in skating, but this did not make her acceptable in her own eyes for the camera. She held up the ceremonies to apply additional makeup, while her exasperated competitor, the silver medalist Nancy Kerrigan, snickered. It highlighted a mean little thought some people harbored: Oksana's performance might have been better, but Nancy looked better.

Men have greater freedom to compete on many playing fields. Athletes, political figures, and CEOs do not fight it out over the Mr. Universe prize. But ask a politician or an athlete about his opponent, and you can see envy as venomous as that of one woman appraising the looks of another. Envy is always focused on competition for prime resources, and, for women, looks play a prime role in their fate in life. As Aristotle Onassis said, "All the money in the world would make no difference if women did not exist." The spoils of success are sexual success, and, for women, this still revolves around the way they look.

Competition among women based on looks is bruising. As Fran Lebowitz has said, most women respond to envy "by being upset, by feeling guilty, by feeling that they've hurt people's feelings. Men recognize envy for what it is—a sign of success. And it spurs them on." Consumer cultures have brought the beauty competition among women to frenzied heights. We don't compare ourselves to people whose attainments are out of reach, but the idea of the level playing ground is that anything and anyone are within reach. As Bertrand Russell said, "Envy is the basis of democracy." By making all seem attainable, it puts many people in a craving state that is ultimately impossible to satisfy. Russell writes, "If you desire glory, you may envy Napoleon. But Napoleon envied Caesar, Caesar envied Alexander, and Alexander, I dare say, envied Hercules, who never existed. You cannot therefore get away from envy by means of success alone, for there will always be in history or legend some person even more successful than you are."

Today the average woman compares her genetic physical endowments with a few hand-picked models. Despite their surreal beauty, the media insist that their beauty is attainable through hard work and effort and buying the right product. At one time we envied only our neighbors because they were all the world we knew. It must have been a little more comforting: it's one thing to win the neighborhood beauty contest, quite another to be put up against the top one percent in the world.

The Biology of Beauty

Why do both sexes care so much about the looks of their lovers, and why do men seem to care more? The answer is sex. The biological purpose of sex is reproduction, not fun or friendship or the communing of souls. For almost all of human history, every mating with a fertile opposite sex partner had some probability of producing offspring. Sex could permanently change the world, bringing forth a new human being who carried inside her the genes of both parents.

Our ancestors' bodies solved the adaptive problem of how to signal their suitability as potential mates. These biological signals are different from the gestures of courtship and flirtation that we use to signal actual interest in the activities our body's beauty provokes. The biological signals are easy reads, the psychological signals are more complex. But if our ancestors did not have radar for healthy, fertile bodies we'd have become biological dead ends long ago. As Charles Darwin said of romance and attraction, its "final aim is really of more importance than all other ends in human life." He meant the composition of the next generation, the survival of the species, or, as a modern evolutionist would say, the survival of the genes.

Humans, like flowers and animals, inhabit a form that is both functional and aesthetic. One exquisite orchid mimics the form of a female wasp, and the bedazzled male wasp makes a confused attempt to copulate. White flowers are often heavily scented (think of perfumes made from jasmine and gardenia) and are most fragrant at night. They are both light in color and nocturnally aromatic because they are pollinated by nocturnal insects. Each species tailors its display to its intended audience. In the nineteenth century some theologians believed that flowers were made beautiful as a divine gift for man's enjoyment; now we recognize this as absurd. Ribbons of tropical fish in Caribbean waters are displaying for each other, not for the humans snorkeling in their waters.

Human sexual displays are specifically designed to inflame our desires, and therein lies the secret of their unnerving power. Beautiful human features are a language, devoted to the adaptive problem of how to visually signal one's own value as a potential mate and how to assess the mate value of others through their visuals. It may seem preposterous that obsession with human beauty is, at rock bottom, an evolutionary adaptation for evaluating others as potential producers of our child. But this is looking at human nature as it would have been designed in the late twentieth century, when sex and reproduction have partially gone separate ways.

As writer Joyce Winer observed after years of birth control followed by fertility treatments, "If the sixties was all about sex without babies, the nineties is all about babies without sex." Sex is more often a source of pleasure than a means to the end of making a baby, and babies are now conceived on doctors' tables and at home with a woman and her turkey baster. But even if sex always and only produced babies, we would still not be having sex or wanting sex with the vast majority of people we see and meet in a lifetime. But our minds are products of a world that was tribal, not global, where birth was not controlled, where the average number of years of life was forty, not seventy, and where infants and children often died before reaching maturity. A biological system that automatically scanned everyone for sexual viability was adaptive. Today, we're left with sneaky feelings for strangers and sexual reactions to faces and bodies we may wish we could evaluate more neutrally.

All people mirror the same form in infinite variations. Our skin varies in pore size, tendencies to be hirsute or baby smooth, blemished or spotless, taut or slack. We have hair that is fine or coarse in texture, abundant or sparse. Our bodies have a vertical height, a distribution of body fat, symmetry or asymmetry of features, and minor variations in the size and shape of our facial features. Men and women differ in their size, the shape of their bodies, the bone structures of their faces, and, of course, their sexual organs. These are the raw materials that selection can act upon. We don't form friendships with people based on the curve of their lips or the slimness of their waists, but we might be attracted to someone because of such a physical detail. We notice even the subtlest differences in human physical appearance and tune in to them with keen interest when we zero in on someone as a potential partner.

Fertility Goddess

The male gaze has been described as target seeking. Psychiatrist Robert Stoller has described "most men of most cultures" as "a whole race of erotic minifetishists." Men talk about being a "leg man" or preferring big breasts or buttocks, and can go into extraordinary detail about what body parts they like. Evolutionary psychologists suggest that men are automatically excited by signs of a woman who is fertile, healthy, and hasn't been pregnant before.

Why would a man care about a woman's nulliparous status (never having had a baby)? Donald Symons believes that there are two reasons. One is the male proprietary interest in female fertility—he wants to be the father of all her babies. The second has to do with the relation between fertility and first birth. For the ninety-nine percent of human history in which birth control was not available, women were usually continuously pregnant or breast feeding after giving birth for the first time. Donald Symons and Margie Profet calculated that this would mean that she was probably infertile ninety-nine percent of the time. This is how they arrived at that estimation: if this hypothetical woman had the usual career of making babies from age sixteen to forty-two, she would have spent six years pregnant and eighteen years breastfeeding. Since breast feeding inhibits ovulation, this would leave her with only twenty-six ovulatory cycles of three fertile days each, meaning that she was capable of conceiving on 78 of 8,030 days or one percent of the time. Of course, this may be an overstatement, since breast feeding may suppress ovulation for months rather than years in younger women. Still, as Symons suggests, until the dawn of birth control in this century, the best way to find a fertile woman was to grab her young and before she started making babies. Perhaps this is why men often prefer the physical signs of a woman below peak fertility (under age twenty). It's like signing the contract a year before you want to start the job.

Female animals stay fertile until they die, but human females do not, and signs of age are therefore important cues to reproductive capacity. A woman's peak fertility is between the ages of twenty and twenty-four and remains near peak throughout her twenties. By the end of her thirties a woman's fertility has declined by thirty-one percent, and after that fertility declines much more steeply. Usually by her early fifties a woman has reached menopause. Things are very different for males, who can still father a baby naturally at age ninety-four. There is no visible sign of a good sperm carrier, or at least no one has found one. Unlike with women, fertility is not written on the male body. This difference is the sole basis for the erotic visual preference for women in their teens and twenties.

Menopause is a cruel biological limitation for a woman who wants to conceive her first child at forty rather than sixteen. It is more than a minor annoyance to the woman who does not want to conceive but who wants to appeal to men as much as she did in her twenties and early thirties. Women do not want to undergo what Susan Sontag has called the "miniature ordeal" of stating their age to men and even to other women who hold "a double standard of aging," making her feel that her years on earth are a source of humiliation rather than pride. Model Lauren Hutton summed up the career trajectory in her business: "As soon as they were out of eggs, women were out of business."

But if elephants and tortoises have fertilizable eggs that last for sixty years or longer, humans could have had them too. A mutation could have altered the rate at which eggs die. But this did not happen. Physiologist Jared Diamond believes that the shutdown is biologically strategic, and that women's fertility operates under a "less is more" principle. Human infants have a long period of helpless dependency, and children have another long period in which they are no longer helpless but still dependent. At some point the woman risks more by having more babies than she does by not having them. By stopping reproduction, her body helps ensure that her current investments, her children, will have a mother who is alive and can care for them, and has not spread her resources too thin or died in giving birth to another child.

The medical science of fertility and reproduction now makes it possible for women to have babies into their sixties. When they are out of eggs, they can be implanted with someone else's; when their hormonal status is not conducive to childbearing it can be artificially altered. A geneticist recently told me that within the next decade she believes it will become routine for young women to remove and freeze some of their young, healthy eggs so that they can be used for later-life pregnancies.

Have all of these changes altered our tastes in beauty and made age and fertility cues in women obsolete? In a world guided solely by thought, not instinct, the answer would be yes. But we are products of evolution and cannot change instincts as quickly as we can change our tastes or update our information. The frenzy over beauty and the enormous business in mimicking youth show that we are still turned on by the usual suspects. It may be difficult to change human nature, and easier to start by fooling her. With the rise of the physical fitness cult, plastic surgery, and advances in beauty technology, the appearance of a woman in her thirties, forties, and even fifties can mimic that of an ancestral woman in her late teens and twenties. One could say that this mimicry is the goal of the billion-dollar beauty industry. And it has been very successful. The highest-paid U.S. actress in 1996 was Demi Moore, whose virtually nude body was the star of *Striptease.* Demi Moore is the mother of three children with the body of a teenage nullipara.

A man may have no interest in getting a woman pregnant, he may take elaborate precautions not to, but his mate detectors are still firing, and he is still inexplicably turned on by the woman who flashes abundant evidence of her fertility. And women are still imitating the appearance of this visually preferred age group, even if they never want to be pregnant at all.

The Importance of Being Resourceful

Some physical qualities lure the other sex like a bear to honey. Darwin called this distinctly sexual advantage "sexual selection." But beautiful ornaments develop not just to charm the opposite sex with bright colors and lovely songs, but to intimidate rivals and win the intrasex competition—think of huge antlers. When evolutionists talk about the beauty of human males, they often refer more to their weapons of war than their charms, to their antlers rather than their bright colors. In other words, male beauty is thought to have evolved at least partly in response to male appraisal.

Male looks are important in establishing dominance hierarchies among men. Males form ranks quickly, even as boys. In boys' camps, rank order develops in cabins within an hour. The top-ranked boy is not necessarily the biggest, but often the best-looking, most athletic boy who shows the most mature physique. The top boy initiates and organizes, and lower-ranking boys obey and question. Their submissiveness is rewarded by the dominant boy's protection and his leadership.

Later in life, handsome men are still the leaders of the pack. In a study of cadets at West Point, sociologist Allan Mazur found that facial dominance had a big effect on military rank in the junior or senior years. Men who looked dominant were handsome and had prominent as opposed to weak chins, heavy brow ridges, deep-set eyes, and ears close to their heads. Their faces were wide or rectangular. Submissive-looking men had narrow or round faces, and their ears were more likely to stick out and their chins recede. Picture alone could predict rank and attainment at the academy and attainment in later career. However, dominant-looking men who did not have high achievement at the academy ended up faring worst of all. They suffered penalties for being sheep in wolf's clothing—when they failed to show the leadership their faces advertised, they were punished, just as if they had lied about their abilities or potential or submitted a false résumé.

But it isn't enough for strong men to battle each other for the hearts (and bodies) of women: they need to appeal to the women directly. One way they do this is through displays of their status and resources. Ovid wrote about this phenomenon two thousand years ago: "Girls praise a poem, but go for expensive presents. Any illiterate oaf can catch their eye provided he's rich." In David Buss's study, women valued good financial prospects more than men did in thirty-six of thirty-seven cultures (in Spain the difference between the sexes was not significant), and valued qualities such as ambitiousness that lead to financial success in thirty-four of the cultures. As promoter Don King remarked about Mike Tyson, "Any man with forty-two million looks exactly like Clark Gable."

Men with higher-status jobs and higher permanent incomes are more likely to be married than men with lower-income, lower-status jobs. For the already married, the possibility of separation and divorce increases if a man's relative earnings decline (relative to his past earnings, or relative to his peers). When anthropologist Suzanne Frayser surveyed separation and divorce among men and women in forty-eight traditional cultures, she found that "incompatibility" was one of the top two reasons cited by both men and women for why the relationship failed. The other important reason for women was that the man had failed to meet his economic and domestic responsibilities. For men, the number one reason they abandoned a wife was reproductive problems.

The sex differences are driven by differences in biology. Men's role in reproduction can start and finish with a few minutes of sex with a fertile woman. A woman risks pregnancy, childbirth, and a potential lifelong commitment to the baby. A man can fertilize as many women as will let him because his body is constantly replenishing sperm. A woman reproduces one baby at a time, and by one man at a time. No matter how many lovers she has, her body limits the rate of reproduction. Men can leave zero heirs or hundreds, even thousands. The average woman can produce no more than eleven children, whether she has one lover or a thousand. For a woman, quality of mates is more important than quantity, and it pays to consider the long haul (if there is a baby, will he be a good father?) as well as the immediate pleasure.

The asymmetries in the roles of men and women in reproduction, and the complexities of assessing which men have the means and desire to invest in offspring lead the two sexes to different mating strategies from the very first glance. Men respond to pictures of women almost instantaneously, and one man's rating is remarkably like another's. Males focus more on pure physical appearance because appearance gives many clues about whether a woman is healthy and fertile, able to successfully carry off a pregnancy, and whether she may be receptive to that man.

A woman makes her evaluations of men more slowly, and if another woman offers a different opinion, she may change her mind. The longer she is left to stare in private at a photograph of a man she has rated as handsome, the less good-looking he becomes in her eyes. If women take a second look, compare notes with other women, or change their minds after more thought, it is not out of indecisiveness but out of wisdom. Mate choice is not just about fertility—most men are fertile most or all of their lives—but about finding a helpmate to bring up the baby.

Throughout the millennia, resources were acquired by being strong and being the best hunter. So it should come as no surprise that men's looks may not be very important to women, but that they are matters for consideration. For thousands of years women depended on men's ability to bring home meat proteins to supplement the diet of fruits and nuts the women gathered. Men defended women against predators and other men, allowing women to focus on protecting the young. Ancient weapons such as stones and clubs were all wielded by hand, making men's greater upper-body strength an even greater advantage. Men are taller than women, and have denser muscles, less fat, and more upper-body strength. Their lung capacity is higher and their blood has more hemoglobin.

The modern man may lift no more than a pencil or a powerbook but good-looking men are still described in terms emphasizing size. The OED defines "handsome" (when used to describe human appearance) as "having a fine form or figure, usually in conjunction with full size or stateliness." Model Hoyt Richards calls his fellow male models "hunky chunks." The dictionary definition of a hunk is a large lump or piece—or a sexually attractive man with a well-developed physique.

When reporter David Remnick covered the National Basketball Association he noticed that "wherever the players stayed the hotel lobby resembled the waiting room at a modeling agency. The women fairly auditioned for them." Basketball player Dennis Rodman has said, "Fifty percent of life in the NBA is sex. The other fifty percent is money," and Remnick comments that "Rodman is exaggerating only slightly." Professional sports figures, like rock musicians, attract women who actively seek to mate with them. Instead of appealing to women through songs of lust, tenderness, and passion, sports figures show off their bodies and their physical skills, achieve dominance over other men and, like famous musicians, make a lot of money. Basketball players have another key advantage: they are very tall and women find tall men attractive.

Despite claims that men's beauty is ageless, it appears that there is a visually preferred age range for male beauty. Donald Symons has conjectured that males in their late twenties reach the height of their physical beauty. A man's natural (untrained) strength increases steadily until age twenty-five to thirty. Male models and strippers tend to be in their late twenties. In the traditional medieval division of life, for example in the writings of Isidore of Seville, it is at twenty-eight that a man has achieved his greatest strength, intelligence, virtue, and physical beauty. In 1500, Albrecht Dürer painted himself at the age of twenty-eight in his most famous self-portrait in which he depicts himself in the image of Christ. Dürer, one of the principal authors of Renaissance canons, painted himself at the moment when he believed he had achieved physical perfection.

But a man's physical beauty can be challenged by another man's status. Henry Kissinger said, "Power is an aphrodisiac." His short, stocky, bespectacled figure beside his taller, younger, and more attractive wife testified to the male-power-female-pulchritude nexus, and suggested to men that they can get any woman they want if they are powerful enough. Even in the animal world females can be lured by males with big gifts and good territories. Female scorpionflies will not even look at a male unless his gift, a tasty bit of insect protein, is at least sixteen square millimeters. If she accepts his gift, they copulate, and the copulation lasts as long as the meal does. Better gifts lead to longer copulations, which are doubly advantageous because they are more likely to lead to conception, and because the extra protein the female is eating will produce better eggs. Of course, even here, looks are not a matter of indifference since the size of the male's gift correlates with his hunting abilities and therefore his physical prowess.

Anthropologist John Marshall Townsend showed people pictures of men and women who ranged from great-looking to below average and who were described as training to be in either low-, medium-, or high-paying professions (waiter, teacher, or doctor). They were asked whether this was a person they might like to have a cup of coffee with, date, have sex with, or even marry. Not surprisingly, women preferred the best-looking man with the most money. But below him, average-looking or even unattractive doctors received the same ratings as very attractive teachers. Status compensated for looks. This was not true when men evaluated women. Unattractive women were not preferred, no matter what their status.

When Townsend and his colleague Gary Levy made the men's status differences more extreme, the women's preferences were even clearer. Townsend and Levy took photographs of men wearing either a Burger King uniform and a baseball cap or a shirt, tie, and blazer, and a Rolex watch, and showed them to women. Some women saw Tom and Harry wearing the Burger King outfit and Jim and Dan wearing the suit and Rolex, others saw Tom and Harry with the suit and Rolex and Jim and Dan in the Burger King outfit. Women were unwilling to date, have sex with, or marry the men in the Burger King outfit but were willing to consider any of these when he was wearing the suit and Rolex. It was an interesting demonstration of "clothes make the man," or that *emblems* of income and status make the man.

Women don't live in bubbles but in families, and in a culture. If a woman marries a man who makes a smaller income than she does, others will say that she has "married down" no matter what his other assets may be. A female CEO for a Fortune 500 company might feel very attracted to her personal trainer or the waiter who serves her lunch, but she faces the problem of introducing him into her world and dealing with his feelings. As we will see, some men find women who make more money than they do less attractive. Simply put, there are social forces that make both sexes uncomfortable in such situations. Men are evaluated by their income and professional status as harshly as women are evaluated by their looks. As feminist author Letty Cottin Pogrebin has written, "Class is the deep dark secret of society, and class plus gender is as volatile as you can get."

Category Crisis

If women had the same access to resources that men have, would the two sexes be more similar in their mating preferences and priorities? Perhaps if the world's economic and political powers were more evenly dispersed we would see a radical reworking of sexual preferences. It is still a world where, as the Humphrey Institute of Public Affairs found, "Women represent fifty percent of the world's population, they perform nearly two-thirds of all working hours, receive only one-tenth of the world income and own less than one percent of world property." A world of equality exists largely in our imaginations and aspirations. As Fran Lebowitz quips, "Men own the joint."

But the evidence so far is that women with high incomes or with financial security want men with even greater assets. When Jacqueline Bouvier Kennedy married a second time, it was to billionaire Aristotle Onassis, and Princess Diana's post-Charles romance was with Dodi al-Fayed, the scion of the billionaire owner of Harrods. Female medical students who expect to pull down large salaries say that they want to marry men whose incomes are equal to or higher than their own: not a single one reports wanting to marry a man who makes a lower income. In surveys of college students, women whose expectations for earnings are the highest place more importance on their future mates' financial prospects than do women who expect to earn less.

Women's wealth has a less clear impact on marriage. In the same study that showed that women medical students want mates who make the same or higher income, sixty percent of the male medical students said that they preferred a mate who made *less* money than they did and forty percent preferred a spouse whose occupational status was *lower* than theirs. Whoever wants less of anything?

Call it sexual blackmail. Even the most brilliant of men have warned women away from pursuits where they might compete on common ground. Immanuel Kant wrote in the eighteenth century that "laborious learning or painful pondering even if a woman should succeed in it destroys the merits that are proper to her sex . . . they make of her an object of cold admiration but at the same time they will weaken the charms with which she exercises her great power over the other sex."

Men hoard resources. They battle with other men and they fight just as fiercely against women. One reason is provided by anthropologist Helen Fisher, who demonstrated that divorce rates soar when women achieve economic independence. This happens in tribal as well as capitalist cultures, in poor countries as well as rich ones. This has several interpretations, one being that women with greater earning power have less to lose by divorcing and are emboldened to do so.

Economist Gary Becker applies the logic of "gain from trades" to explain the female migration. If Japan makes better televisions and the United States makes better airplanes, a trade will be advantageous to both. If Japan begins to make its own high-quality airplanes, the gains from trade will diminish. Women still do the majority of the housework and childcare even when they work full time. The woman who hunts *and* gathers may feel less need of a fellow hunter for whom she must still gather.

The newest data on women and work suggest that, while money brings women independence, it does not necessarily make marriage undesirable to them. As sociologist Megan Sweeney writes, it just "alters the nature of the marital bargain." There has been a striking change just over the past few decades. In the 1960s and 1970s women with high-status jobs and high incomes were less likely to marry or remarry than women who did not work or who had lower-paying or lower-status jobs. The opposite was true in the 1980s. In fact, the cover story of the June 15, 1998, *New York* magazine says that, if a Wall Street millionaire leaves his wife to remarry, the new "trophy" wife will be a high-powered female peer. It is likely that eventually female income will be a status symbol in a spouse. This is not to say that the men will not want to be making *more* money, or that the women will not still be seeking men who make even more than they do.

Marjorie Garber, a professor of English at Harvard, suggests that we are now in "category crisis," a blurring of cultural, social, and aesthetic distinctions that mirror a society in transition. Women are flaunting bodies that are taut and toned rather than soft and round. A decade or so ago, a tall muscular female body was considered freakish. Big and buxom or small and delicate, the female physique looked soft and exuded a powerful mix of sexuality and vulnerability. Today the women other women find most beautiful are as fat-free and as tall (five feet ten inches) as the average man. A top female cosmetics company, MAC, has employed two unlikely spokespeople: the drag queen RuPaul and the singer K. D. Lang, who is a lesbian. Lipstick, the symbol of female adornment for male favor, is now being marketed by a man dressed as a woman and by a woman who dresses as a man.

Men are now spending nine and a half billion dollars a year on plastic surgery, cosmetics, fitness equipment, and hair products, including dyes, weaves, and transplants. Some say that they are trying to look young to remain competitive in the workplace, but their motivation is likely to be sexual competition as well. As women spend more time out in the world mingling with their mates' competitors, men may feel the need to add another weapon to the arsenal: their own physical appeal.

Doubtless, confusion and edgy dissatisfaction mark the relations between the sexes. The divorce rate is high, the level of single parenthood is high, and the level of childlessness is the highest it's been since the Depression. But evolutionary psychologists would suggest that instincts that have worked for millennia will be very hard to stamp out. The automatic desire to stare at and desire young female beauty and the sexual attraction of a tall man with a broad chest, a chiseled profile, and money in his wallet will not soon become things of the past. It's not clear we would even want them to: what we want is to be more aware of the choices we make and the forces that impel us, so that we are able to work in our own interests, and not solely in the interests of our genes. Genes don't care about human happiness: but people do, and the person best able to carry forth our genes may or may not be the person we would most enjoy spending our days with.

Money

Beauty is a big advantage in the bedroom, but it is also an advantage in the boardroom. Although it does not approach racism or sexism in magnitude, lookism appears to be a form of discrimination in the workplace. And a silent one. No one thinks that he has been offered a lower salary because he is short! Good-looking men are more likely to get hired, at a higher salary, and to be promoted faster than unattractive men.

For women, the relation between appearance and success at work is less straightforward. Good-looking women, like good-looking men, are more likely to be hired and to receive higher salaries. But this is not always true. A few studies show that good-looking women actually fare worse than plainer women and that good looks in a woman can "backfire." In one study, good-looking women were less likely to be made a partner in a law firm, in another they were less likely to be hired for managerial positions. One explanation is that good-looking people of either sex get more "sex typed." Good-looking men look masculine and good-looking women look very feminine (in their facial features and bodies, not necessarily their clothes or style). This helps a man, who is assumed to be powerful, in the know, and independent, but it hurts women, whose appearance brings the unwarranted assumption that they may be submissive and overly sexual rather than tough and decisive. However women's looks are "read," they don't advertise the kind of qualities generally sought by management teams.

In one well-known study out of Columbia University Business School conducted in 1979, Madeline Heilman and Lois Saruwatari found that good looks helped a woman to be hired (and hired at a better salary) for a clerical job, but worked against her in being hired for a managerial-level job. Later studies suggest that women are rewarded for their beauty in any job with high visibility where interpersonal skills are required but they are not rewarded and may even be penalized in jobs calling for the ability to work well under pressure, make snap decisions, and motivate others. Heilman and Saruwatari conclude, "This finding sadly implies that women should strive to appear as unattractive and as masculine as possible if they are to succeed in advancing their careers by moving into powerful organizational positions. Surely giving up one's womanhood should not be a prerequisite for organizational success."

This study appeared in the same decade when John Molloy published his best-selling *Dress for Success* primers positing that success in business and sexually attractive attire were almost mutually exclusive. The *Dress for Success* edition for women was on the best-seller list for five months. However, by 1987 *Mademoiselle* magazine was already reporting that its influence was on the wane. Yet as women have reclaimed the right to look like women instead of imitation men at work, sexual tensions have escalated. Good-looking women are the most likely to be sexually harassed by male coworkers, who often underestimate just how unacceptable sexual aggression is to women. For example, asked how they would feel if they knew a coworker wanted to have sex with them, sixty-four percent of women said "insulted," sixty-seven percent of men said "flattered." As we saw in the last chapter, men are also more likely than women to interpret friendly gestures as sexual invitations, particularly when the gestures are made by attractive women.

The economics of being female are never particularly advantageous. A good-looking woman with high career aspirations may be penalized for her looks because she is assumed to be "too feminine" to do a high-powered job efficiently. She may be sexually harassed by men, and envied and left out by other women. But it is homely women who are truly disadvantaged economically—they are less likely to get hired or to earn competitive salaries at work. They are less likely to marry, and less likely if they do marry to marry a man with resources. These facts alone drive high consumption of beauty products. It may not always pay to look great but it pays to look average.

Happiness

After all we have discussed in the past two chapters, you would have to assume that beautiful people are happier than other people. As Ben Franklin said, "Human felicity is produced not so much by great pieces of good fortune that seldom happen as by the little advantages that occur every day." As we have seen, great-looking people are afforded those little advantages all of their lives, so they *must* be happier.

Beauty, in fact, does not bring much extra in the way of happiness. Psychologists Ed Diener and David Myers have spent a lot of time trying to understand what makes people happy. They focus on "subjective well-being," a state of mind in which a person feels very positive, seldom feels negative, and has an overall sense of satisfaction with life. Ed Diener finds that good-looking men have a somewhat greater sense of well-being and feel a bit happier than other men. A woman's beauty sometimes makes her a bit happier than other women, but it can also make her more unhappy. The overall effect for both sexes is marginal. The biggest effect is on satisfaction with one's romantic life. Here the good-looking are happier. But somehow this does not lead to greater overall life satisfaction.

Why doesn't beauty, that brings so many advantages, bring more happiness? Diener and Myers believe that happiness has more to do with personal qualities such as optimism, a sense of personal control, self-esteem, ability to tolerate frustration, and feelings of comfort with and affection for people than with looks or money. They note that it is human nature to keep adjusting expectations according to circumstances—the more we get, the more we want since we are always comparing ourselves with people who have more. As psychologist Timothy Miller observes, "No instinct tells us that we have accumulated *enough* status, wealth, or love. . . . To the contrary—such an instinctive mechanism would contradict the basic principles of evolution." The good-looking compare themselves with the even better-looking, the rich with the even richer. Automatically running after what you don't have (yet) may give you a competitive edge, but taken to unreasonable extremes, it can lead to lack of self-acceptance and lack of joy. The key to happiness is being able to occasionally override the more-is-better attitude and appreciate and feel gratitude for what you have.

Desire is unquenchable. The psychoanalyst Edith Jacobson has written about beautiful female patients isolated by their beauty. Catered to all of their lives, they become convinced that they can get whatever they want and whomever they want, a stance bound to lead to frustration at each rebuff and setback. As Betrand Russell wrote, "He forgets that to be without some of the things you want is an indispensable part of happiness."

Studies of twins suggest that happiness may be partly under the control of the genes. Behavioral geneticist David Lykken studied fifteen hundred pairs of twins, comparing identical twins who share one hundred percent of their genes to fraternal twins, who are no more similar genetically than other siblings. Lykken and coauthor Auke Tellegen concluded that people are born with a "set point" for happiness, an equilibrium point to which their mood returns after brief fluctuations. In other words, some people will have natural tendencies to worry or brood while others will be sanguine. On a recent episode of the Charlie Rose show, the host chided actor Liam Neeson for not "being on top of the world. How," he asked, "could you not be ecstatically happy, given your career success, your marriage, your life?" Neeson did not say he was unhappy, but just that he was a worrier. The many happy turns in his life had not changed that.

And there is self-esteem, one ingredient of happiness that is more tightly linked to how we see ourselves than to how others see us. As Eleanor Roosevelt remarked, "No one can make you feel inferior without your consent." Our beauty as others judge it is linked to social ease, but it is not linked strongly to self-esteem. Even if others think we are beautiful, we may not if we are constantly comparing ourselves to the even more beautiful. But our beauty as *we* see it *is* linked to self-esteem. Ed Diener speculates that "it seems plausible that happier people tend to perceive themselves as somewhat more attractive than objective ratings might indicate." Happier individuals also enhance their appearance more with clothing, makeup, jewelry, and so on than do unhappy people, thereby maximizing their assets.

Beauty has a downside. People assume that the beautiful may make less faithful partners and may be more likely to seek a divorce. Beautiful women may be seen as less likely to make good mothers, and beautiful men may get questioned about their sexual orientation, no matter what their preference. And beauty can be damn distracting. William Butler Yeats apologized to Anne Gregory: "Only God, my dear, could love you for yourself alone, and not your yellow hair."

When people judge integrity, sensitivity, and concern for others from facial appearance, beauty has little power. A face radiating kindness and sympathy may not be beautiful, and a beautiful face may look aloof, blank, haughty, or self-absorbed without losing its beauty. As Montaigne said, "There are propitious physiognomies; and in a crowd of enemies all unknown to you, you will immediately pick one rather than another to whom you surrender and to whom you will entrust your life and not precisely from considerations of beauty." But even Montaigne concludes, "A face is a poor guarantee; nevertheless it deserves some consideration." Beauty may bring small advantages, even here.

But the downsides are not inconsiderable, particularly for a woman. She may be favored in a million small ways but if what's important to her is to be seen as a good mother, to succeed in a high-level profession, and to be honored for her kindness and integrity, beauty may either be irrelevant or it may even interfere with her chances to be seen as she is, and wants to be. Beauty is not a sure road to happiness.

Despite all this, no one offered a chance to be more beautiful would turn it down. As vaudeville star Sophie Tucker once said, "I've been poor and I've been rich and rich is better."

CHAPTER FOUR

Cover Me

And I have known the arms already, known them all—
Arms that are braceleted and white and bare
(But in the lamplight, downed with light brown hair!)

—T. S. ELIOT

Clothes are a social statement; makeup is closer to our secret
hopes and fears.

—KENNEDY FRASER

You'd see women in the Laundromat with no makeup, a
housedress, bedroom slippers, and a hairdo that'd be appropriate
for the Presidential Inauguration.

—JOHN WATERS

Sigmund Freud said that seeing is "ultimately derived from touching." Nowhere is Freud's observation better realized than at the body's boundary. When journalist Kennedy Fraser spies a woman's pair of long gloves at a glittering social gathering, they look to her like "steamrollered silk arms"—surfaces so smooth and taut, they mimic perfect skin. But such pleasures pale in comparison to the real thing. Flawless skin is the most universally desired human feature, according to zoologist Desmond Morris, and flowing, healthy hair runs close behind.

Skin may be the body's most aesthetic organ; it is certainly its most pervasive. Thickest on the soles of the feet and thinnest on the eyelids, it weighs six pounds and measures twenty square feet in the average person. Within each inch of skin are sweat glands, oil glands, hairs, blood vessels, and nerve endings through which we shiver, shudder, sweat, blush, and quiver. On the skin's surface is keratin, the protein that rhinoceros horns and animal claws are made of, and the stuff of human hair, which is just a special form of skin.

Skin and hair, so sexy and glorious when healthy, are repellent when not. In his writings on disgust, William Miller observes, "There is nothing quite like skin gone bad; it is in fact the marrings of skin which make up much of the substance of the ugly and monstrous. . . . Pus, running sores, skin lesions, which were a regular feature of medieval life and helped define the pariah status of lepers and syphilitics, have only recently come to be rare sights in the West."

Hair is alluring, but only on the head. We are disgusted to find so much as a single hair in the center of a mole, on a woman's chin, or floating in a glass of water. It is said that John Ruskin never consummated his marriage, so appalled was he to find that his wife did not look like the Greek statue he imagined, but came with fleece between her legs. To some, hairy armpits or pubic hair, sprouting at puberty and imbued with pungent odors, disturb. To others, they excite: in the course of their heated affair, Caroline Lamb, the wife of British Prime Minister Lord Melbourne, sent Lord Byron strands of her pubic hair.

Freud believed that we fetishize hair not only because of the way it looks and feels but because of the way it smells. "Both the feet and the hair are objects with a strong smell which have been exalted into fetishes after the olfactory sensation has become unpleasurable and abandoned." What has been abandoned, according to Freud, is "coprophilic pleasure." In other words, our love of the body's smells is transferred from the toilet to the toilette and the coiffure. Skin and hair stir visual, tactile, and olfactory sensations and memories; they are polymorphically arousing and primal in their appeal.

Naked

Desmond Morris called man "the naked ape," the only one among "one hundred and ninety-three living species of monkeys and apes" not covered with hair. It begs the question: why do we show bare skin rather than fur? In fact, we are not entirely naked. We have five million hairs on our adult bodies and as many hair follicles as an ape, but most of our body hair is so fine that we look bare skinned. To stay warm, we've sheered the fleece, tanned the skin, and captured the fur of other animals—in other words, we've invented clothes. We also have a thick layer of fat underneath the skin which, like the blubber of a whale, helps insulate us against the cold.

Our cooling system is equally distinctive. When most animals get hot, they pant, they sweat a little in hairless areas such as the pads of their paws, and they fluff their fur. Humans get sweaty all over. Millions of sweat glands called eccrine glands embedded in the skin turn on like sprinklers when the temperature gets hot. When dry air moves across wet skin, the water evaporates and the temperature of the blood in the capillaries cools down. It is an evaporative cooling system that evolved in the African savanna where the temperature is hot, the air is dry, and the landscape is dotted with water sources.

By becoming downy rather than furry, we've made our bodies a less than desirable home for fleas, lice, mites, and other parasites. Our dogs and cats need to wear flea and tick collars in the heat, not us. Nakedness has erotic advantages, and not just in keeping us parasite-free. Hair offers padding and protection. Strip it away, and the skin is more sensitive and responsive. In areas rich with nerve endings such as the lips, the palms of our hands, the soles of our feet, our nipples, and parts of our genitals, we have no hair at all.

The anthropologist Marvin Harris suggests that our denuding took place when our ancestors got up on two feet and started running long distances. Humans cannot survive by speed alone; we survive on stamina. Our most talented athletes are lucky to equal the speed of a wild turkey in short sprints (about twenty-six miles an hour). Compare that with the speed of thoroughbred horses and greyhounds, which can run at over forty miles an hour, or cheetahs, which can reach speeds of seventy miles per hour. But what humans lack in speed they make up for in endurance. Our ancestors succeeded in the hunt by outdistancing their prey and simply exhausting them. Naked skin cooled by millions of eccrine glands kept them cool for the long run in the sun.

Our skin reflects the fact that we evolved as runners in a hot climate. How much skin we have, relative to our overall size, reflects just how far we have strayed from the environment of origin. In certain parts of the world most people are short and spherical, and in other parts they are tall and cylindrical. The more skin we have relative to our body size, the faster we cool off. This is why, no matter what our size, we like to curl up when it is cold and stretch out flat when it is hot. People living in hot dry environments such as the Dinka in the Sudan are slim, with narrow torsos and very long limbs. This gives them a high ratio of surface (skin) to body mass, and allows their bodies to dissipate heat. Alek Wek was born into the Dinka tribe and is a typically narrow-torsoed Sudanese native who stands five feet, eleven inches with mile-high legs. She is currently gracing magazine covers all over the world as a top model, her body shape the exact proportions of today's tall, slender beauty ideal.

The chunky, short-limbed shape of some Northern people such as the Greenland Inuit appears to be a more recent variant of the human form, emerging in the last few tens of thousands of years as an adaptation to very cold climates. Such bodies, with more mass than surface, conserve heat. All over the world there is a positive relation between the average yearly temperature and length of the limbs—the hotter and drier it gets, the less volume in the torso and the longer the legs and arms.

Groomed

Primates carefully pick parasites and dirt from each other's pelage. Mutual grooming has functioned over millions of years as mutual doctoring and a means of social bonding. When rival male chimps reconcile, they circle each other tensely, shriek and embrace, and then formalize the truce by mutual grooming. Rhesus monkeys crowded into tight spaces groom each other more than usual. Ethologist Franz de Waal observed a young rhesus female he named Azalea who was born with a chromosomal abnormality akin to Downs syndrome in humans. Azalea, who could not have survived on her own, was carried around by her sister well beyond the normal age, and groomed twice as much as her peers.

Mothers of many species groom and lick their pups immediately after birth, and this stimulation appears to have lifelong consequences. When neurologist Saul Schanberg took rat pups away from their mother even for brief periods of time, the pups produced less growth hormone and less ODC enzymes (ornithine decarboxylase activity), which are involved in the timing and initiation of important chemical changes in the body. The maternally deprived rats returned to normal only when the scientists stroked them with a paintbrush. Pups licked and groomed more by their mothers during the first ten days of life have lower levels of stress hormones than other pups. Scientists believe that the mother's licking and grooming programs her pup's biological responses to threat, and regulates its physiology and central nervous system.

Because of their need for intensive care in the first few days of life, prematurely born human infants often have less skin-to-skin contact than other newborns. Psychologist Tiffany Field and her colleagues at the Touch Research Institute at the University of Miami School of Medicine have found that giving premature human infants daily massages has profound effects on their growth and development. The massaged infants gain weight as much as fifty percent faster than unmassaged babies and are more alert and healthier. Early touch and the manipulation of the skin by stroking, bathing, and so on may have important biological consequences for humans as well as animals.

All of our lives we make a ritual out of daily grooming, and we have created a whole industry of professionals to help us—hairdressers, barbers, manicurists, pedicurists, and so on. Parents groom their children, and little girls love to groom their dolls. In her memoirs, Mary Catherine Bateson recalls that her "most intimate time of the day" with her mother, Margaret Mead, came when her mother sat "on a special stool at the foot of my bed" and combed her waist-length hair. Dolls don't sell well unless the child can indulge in what toy manufacturers call "hairplay," coifing and combing the doll's hair. The makers of the most popular doll of all times, Barbie, produced a special "totally hair" version in 1992; her hair reached to her toes. It became the best-selling Barbie ever.

Throughout life, people lavish attention on their skin and hair. In the United States we spend more than twice as much money on personal care products and services as on reading material. Worldwide, cosmetics and toiletries are a forty-five-billion-dollar industry, with North America representing 30 percent of the global market

(Europe represents 34.9 percent of the market, Japan 18.9 percent, and the other countries a shared 16.2 percent). In 1996, 88 percent of women in the United States over age eighteen said they had used color cosmetics in the past six months.

The Food and Drug Administration defines cosmetics as whatever is "rubbed, poured, sprinkled or sprayed on, introduced into or otherwise applied to the human body for cleansing, beautifying, promoting attractiveness or altering the appearance without affecting the body's structure or functions." Is our lavish spending on these products a sign of too much time and money, or of the exploitation of the insecure by media advertising? Unlikely: the practice dates back at least forty thousand years.

Rubbed, Poured, Sprinkled, Sprayed

In the Klasies River Mouth and Border Cave region of southern Africa, archaeologists have found sticks of red ochre that are about forty thousand years old. The sticks were formed by grinding and mixing iron oxides with animal fat or vegetable oil, then heating them to intensify the color. It is unclear what people did with the red ochre, but anthropologist Steven Mithen believes that the crayons were used to paint the face and body, since no known art has been discovered in South Africa from before thirty thousand years ago.

Makeup was an advanced art by the time of the ancient Egyptians. When archaeologists opened the tomb of King Tutankhamen they found a pot of three-thousand-year-old moisturizer made from animal fat and perfumed resin. In the British Museum there is a woman's cosmetics box, also from Egypt, dating to about 1400 B.C. Inside it are an ivory comb, pumice stones, containers of makeup, vases for skin salves, a pair of gazelle-skin sandals, and small red cushions. Shaving sets from 2000 B.C. containing bronze razors and tweezers have been found. Ancient Egyptians stored moisturizers made from animal fat, olive oil, nut oils, seeds, and flowers in jars of alabaster and onyx. Their medical papyri have formulas and recipes for preventing wrinkles and blemishes. Both sexes pumiced and shaved their bodies, and wore wigs, sometimes in combination with their own hair (a look later adopted by Andy Warhol). In fact, the Egyptians had most of the cosmetics we have today, suggesting that the cosmetic business is hardly a modern invention or response to current cultural pressures.

Hair Today, Gone Tomorrow

We pay the most obsessive attention to our face, and comparatively little to the skin of our bodies. But one thing we do is remove hair. Women have less body hair than men, but in many cultures they are the ones to remove what they have. By doing so, they accentuate the difference between the male and the female body, and heighten their appeal to men. The Roman poet Ovid advised women, "Let no rude goat find his way beneath your arms and let not your legs be rough with bristling hair." Supermodels don't walk the runway with hairy legs. Their skin looks as smooth and molded as that of plastic dolls.

Although greater naturalism has pervaded images of female bodies in both high art and smut, body hair is still, at best, a sometime thing. *Playboy* models used to have their pubic region hidden or deforested. In current issues, the models show perhaps a bit of pubic fuzz. Botticelli portrayed women without body hair, but centuries later, so did Degas, Matisse, and Picasso. At the fashionable Frederic Fekkai salon in New York, hundreds of women request waxes of every hairy part of their bodies including pubic waxes, which range from a trimmed triangle to a vertical one-inch strip on the pubic mound (a mohawk), to removal of all hair. The trend began as a bikini wax, to eliminate hairs which strayed outside the confines of a bathing suit, but has gone to the fashion extreme.

Whenever a new part of the female body is bared, it becomes a potential site for hair removal. In the fourteenth and fifteenth centuries some European women covered their hair and ears behind wimples. The small bit of hairline visible where the wimple met the forehead was soon plucked, and then plucked some more. What resulted was the fashion for high broad foreheads, receding hairlines, and plucked eyebrows, seen in Renaissance portraits by Jan Van Eyck and Rogier van der Weyden.

Men in Western culture are more likely to hide all parts of their bodies beneath clothing. But when they bare their chests, attention is paid to the look and feel of the skin. Body builders apply "competition color" (fake tans) and body oil. They also shave their torsos, so that hair will not interfere with the display of defined muscles. Clive James described the torso of a body builder as a "condom filled with walnuts." There has been only one hairy-chested Tarzan, Mike Henry, among all the screen Tarzans. Male supermodel Marcus Schenkenberg's muscular, tanned, and hairless chest adorns men's clothing advertisements all over the world. Some gay men go in for body waxing so that their torsos conform to what Michelangelo Signorile calls "the shaved muscle boy aesthetic." As Blanche DuBois crooned in *A Streetcar Named Desire*, "I like my men smooth and hairless."

But their goal is not to mimic the hairlessness of a woman; no matter how much body shaving or waxing some men may do, they don't wax their faces. They often appear in formal shots and on magazine covers with a bit of manly stubble. Shirtless men often pose with fingers looped through tugged-down pants or wearing baggy pants which fall low. When the skin below the navel is exposed it is never shaved, nor are the arms or legs. On the chest, hairs are plucked to give the torso the look of armor. The skin is the vestment of the gladiator, and the chest becomes a bronzed carapace. Hair would simply ruin the effect.

The Decorated Body

Decorated bodies are now in vogue, but until recently the body arts were seldom seen in the West. Even in 1990 an informal survey found that only about three percent of Americans have tattoos, and most of them are men. But clothes get shed, and billboards tout men's bare chests, and slip dresses rule the runway, exposing more and more skin. As the shock value of bared skin is waning, adorned skin, virtually wiped out by pious Westerners in many parts of the world, is enjoying a renaissance.

Tattooing is believed to have originated in Nubia, in the fourth century B.C. The word derives from a Tahitian word which means to strike. Tattoos have been made with boars' tusks, sea turtle shells, or fine needles which are used to puncture the skin and inject dyes. In the nineteenth century Darwin found tattooed aborigines "from the polar regions in the north to New Zealand in the south."

Cicatrization, or scarring of the body, is created by raising the skin into patterns with a knife or another instrument and is not uncommon in parts of the world where skin is dark and tattoos are hard to see. Piercing is practiced everywhere. Mummies have been found with elongated earlobes caused by the use of heavy earrings, and with two piercings in a single ear. People throughout human history have shoved shells, bones, feathers, and metallic objects through their ears and noses. They have pierced and bejeweled virtually every part of the face—the ears, noses, lips, eyebrows, and tongue—and all the erogenous zones of the body—navels, nipples, penises, and labias. Piercings in nerve-rich areas stimulate constant sensation, and viewers can't help but imagine these sensations as they look at the pierced skin. As such, the appeal is not so much visual in nature as tactile.

When European missionaries, merchants, and explorers ventured into the world in the eighteenth and nineteenth centuries, they were shocked to find that their mode of dress was not universal. They found people who were not only naked but painted, scarred, and tattooed. Despite the striking difference in appearance of adornments, the paints and scars were actually serving much the same purpose as the clothes and uniforms of the Westerner. As art historian Anne Hollander observed, "The definition of 'dressed' may sometimes be so elastic that the distinction seems quite different from the one we are used to . . . people who do not wear garments nevertheless develop habits of self-adornment that seem, as Western clothing does, to be a necessary sign of full humanity: they are ways of clothing the human body in some completed concept of itself. . . ."

Like clothes, tattoos, scars, and piercing are seen as beautiful by their owners, but beauty may not be their primary purpose. They are emblems that distinguish people from one another by rank, status, gender, age, and accomplishment, and in fact can function as specifically as a uniform. In many cultures a tattoo or scar marks puberty, marriage, first successful hunt, and so on. They are distinctive to a particular tribe.

Tattoos and scars, like clothes, can also take on idiosyncratic meanings, particularly when used in Western cultures. Lovers tattoo the names of the beloved to their arms much as they would carve them into a tree's bark. But tattoos, scars, and piercing carry an additional, important message. In many parts of the world, tattooing and scarring are part of the initiation rites for adolescents. Like smoking and drinking and other forms of bravado, tattoos, scars, and piercings on Western adolescents may serve a similar function. Tattoos and scars are not only painful to receive but carry risks of infection from blood-borne pathogens and thereby advertise the strength and resistance of their recipients. Like hunters and warriors who have worn battle scars proudly, the tattooed, pierced, or scarred individuals are showing that they faced down physical challenge. The body arts are tough.

Face Paint

The body arts are practiced by both sexes, but at least in the West, they have existed more within a masculine domain. But face paint belongs to women. Occasionally, rouge or powder has made its way onto a man's cheeks, but lipstick or eyeliner is and always has been a fringe activity for men. The male vanity market has expanded in recent years but the only segment of the market that has not expanded is men's makeup.

Women conceal, bleach, and blush. They have applied poisonous lead and mercury to their skin mixed in with egg whites, lemon juice, milk, and vinegar. They have attached leeches to themselves and swallowed arsenic wafers. To mimic skin translucency the Greeks and the Romans and, later, Queen Elizabeth I painted the veins on their breasts and foreheads blue. For two thousand years European face makeup was made from white lead, which was combined with chalk or used in a paste with vinegar and egg whites and applied thickly to completely mask the skin's surface and color. The Spanish physician to Pope Julius III, Andreas de Laguna, complained that he could cut off "a curd of cheesecake from either of their cheeks," so thick was women's makeup.

In China and Japan women have used a similar palette of white face paint with red rouge and nail coloring. In the Heian period in Japan, from the ninth to the twelfth century, women applied to their skin a thick chalky powder, called oshiroi, made from rice flour (and later, white lead) and a rouge made from extract of safflower called beni. Japanese women still value very pale skin. At the Clinique counter in Japan, one finds a whitening treatment with sales as brisk as those for cleansers and lipsticks. As Japan's brand manager for Clinique explained, Japanese women believe that good skin "is absolutely fair and uniform. They hate freckles."

Whitish-pink face makeup persevered in the United States until very recently. Makeup artist Kevyn Aucoin relates how, in 1967, as a first grader, he mistook a store clerk's face makeup for calamine lotion. Commiserating with her about his own mosquito bites, he encountered puzzled silence. On the way home, his mother explained that the pale paste was face makeup.

Over this pale canvas, women apply exclamation points of red to their lips and cheeks. Red, the color of blood, of blushes and flushes, of nipples, lips, and genitals awash in sexual excitement, is visible from afar and emotionally arousing. For the same reasons, red is the color of stop signs, railway signals, and fire trucks. Red pigments were applied to the lips in 5000 B.C., placed in cartridges as lipstick in Paris in 1910, and thereafter packaged and sold in so many tubes that by 1930 these tubes could stretch from Chicago to San Francisco. In the United States today, 1,484 tubes of lipstick are sold per minute, and many women feel as designer Betsey Johnson does: "If I were dying, I would be in the hospital wearing lipstick." Pots of red oxide of iron have been found inside ancient Sumerian and Egyptians tombs. Apparently, it's a timeless last request.

Secrets and Lies

Historically, female use of face makeup has been unsettling to men but stubbornly unyielding. In ancient Rome, the poet Martial wrote, "You are but a composition of lies. Whilst you were in Rome, your hair was growing on the banks of the Rhine; at night, when you lay aside your silken robes, you lay aside your teeth as well; and two thirds of your person are locked up in boxes for the night. . . . Thus no man can say, I love you, for you are not what he loves, and no one loves what you are." Ovid warned, "Your artifice should go unsuspected. Who could help but feel disgust at the thick paint on your face melting and running down onto your breasts? Why do I need to know what gives your skin its whiteness?"

Church leaders expressed indignation. St. Jerome wrote, "What makes this purple and white stuffe in the face of a Christian woman, the inflamers of youth, the nourishers of lust, and tokens of an unchaste soule?" Clement of Alexandria declared wig wearers unable to receive the Lord's blessing because the blessing stayed on the wig and did not penetrate to the person. "For on whom does the presbyter lay his hand? Whom does he bless? Not the woman decked out, but another's hairs, and through them another head."

A law was passed by the English Parliament in the late eighteenth century which attempted to impose on women the same penalty for adornment as for witchcraft, freeing up the husbands who had married them under such false pretenses. "All women . . . that shall from and after this act impose upon, seduce or betray into matrimony any of His Majesty's subjects by the use of scents, paints, cosmetics, washes, artificial teeth, false hair, Spanish wool, iron stays, hoops, high heeled shoes, or bolstered hips, shall incur the penalty of the law now in force against witchcraft and like misdemeanors and that the marriage, upon conviction, shall stand null and void." The law was unenforceable.

In 1711 the British journal *The Spectator* published a letter from one distressed husband which read, in part, "Sir, . . . I have a great mind to be rid of my Wife, and hope, when you consider my case, you will be of Opinion I have very just Pretensions to a Divorce . . . never Man was so enamored as I was of her fair forehead, Neck, and Arms as well as the bright Jet of her Hair, but to my great Astonishment I find they were all the Effects of Art: Her skin is so tarnished with this Practice, that when she first wakes in a Morning, she scarce seems young enough to be the Mother of her whom I carried to Bed the Night before. I shall take the Liberty to part with her by the first Opportunity, unless her Father will make her Portion suitable to her real, not her assumed, Countenance."

Kings had painters visit their future matrimonial prospects and bring back portraits. Although these marriages were "arranged" with an eye toward status enhancement (united territories or powerful families), the appearance of the woman was considered important. Henry VII of England gives us a glimpse of what these kings were looking for. In addition to the official portrait, he asked his traveling coterie to answer a lengthy questionnaire about the intended, including instructions such as to "marke the favor of hir visage," to note "whether she bee paynted or not," and "to marke whether there appere any here aboutes hir lippes or not."

Female royals and aristocrats were not expected to take as strong an interest in the appearance of their husbands, but they often asked for and received portraits. Elizabeth I refused to marry anyone she couldn't see. But then, she rejected everyone and died the Virgin

Queen. Until the end, she enjoyed flirting with her suitors, her face coated with white, her cheeks painted with circles of red, her head covered in a wig of golden and red hair. Perhaps unsure how effective this was, she refused to look in the mirror for the last twenty years of her life.

Why Marilyn Became a Blonde and Elvis Dyed His Hair Black

Fair, blushing skin is the skin of youth, of the female, of the woman who has never borne a child. It is why women of all ages have struggled so to maintain it for life. They are trying to mimic the beauty of the nubile adolescent and in so doing join in the universal obsession with clear skin and the various ruses to mimic it. Face paint, and increasingly facial plastic surgery, is all about the illusion of youth and fertility.

In our multicultural society we associate skin color with race. But skin color varies not only with race but with sex and age. In fact, in the small homogeneous populations from which we evolved, sex was the main source of skin color differences. This has nothing to do with sun exposure—it is true for parts of the body not exposed to sunlight, and in cultures where the women spend as much time out of doors as the men. Women tend to be paler than men of the same race because women tend to have less hemoglobin in their blood and less melanin in their skin. Artists in ancient Egypt, Crete, and Japan highlighted the differences between males and females by using different colors in their representations—white, yellow, or gold for women, orange, red, or brown for men. Millennia later, sandy-haired Elvis Presley dyed his hair blue-black and, to be sure everyone noticed, constantly drew his hand back along it; and the medium-toned, freckled brunette Marilyn Monroe wore pale makeup and dyed her hair platinum. To exaggerate their sexual appeal, each heightened signs of their masculinity or femininity through skin and hair color.

Babies have paler and more delicate skin and hair than their parents. Hair never gets blonder and skin never lightens naturally as people get older, and true flaxen hair is seldom seen beyond early childhood except among Scandinavians. Anthropologist Peter Frost has suggested that women's pale skin may serve similar purposes to those of infants: as a deterrent to aggression and as a general signal of youthfulness.

But as he also points out, skin color differences between men and women are products of sex hormones and directly indicate a woman's fertility. Young girls and boys do not differ markedly in skin tones. The dimorphism emerges only at puberty, when boys darken and girls lighten. Thereafter, women are lighter during ovulation than during the infertile days of their cycle. Their skin darkens when on the pill and when pregnant. Pregnancy is advertised not just in changed body shape but in skin changes from dark blotches on the face ("the mask of pregnancy") to darkening of the nipples to spider veins and stretch marks. Although most of these disappear after the baby's birth, a woman's hair and skin tend to be permanently darkened after the first pregnancy, forever changing the girlish complexion of youth. The relative lightness of a woman's skin then is a direct readout of her hormonal state.

The neuroscientist V. S. Ramachandran has suggested yet one more possible reason that selection pressures may have favored lighter skin in women. Light skin is a more transparent window to health, age, and sexual interest than darker skin. In his article "Why Do Gentlemen Prefer Blondes?" he suggests that blondes are preferred not so much for their hair color as for their fair skin, which tends to accompany light hair. According to Ramachandran, fair skin makes it easier to detect signs on the skin of disease (anemia, cyanosis, jaundice, and infection), signs of sexual interest and arousal (blushing and flushing), and signs of aging. Although both sexes are interested in a mate's health, signs of youth are more important to men, and signs of sexual interest harder to read on women, who often signal them less overtly than men. What Ramachandran is suggesting is that light skin was sexually selected by men because light-skinned women were the least able to deceive them. What was biologically advantageous became an aesthetic preference.

Anthropologist Douglas Jones suggests that the male preference for lighter than average skin color in a female partner is a cultural universal or near universal. In the Kama Sutra, written in the first and fourth centuries A.D., the ideal woman, the Padmini or Lotus woman, is described as having skin that is "fine, tender and fair as the yellow lotus." Sociologist Hiroshi Wagatsuma noted that Japanese men value "whiteness of skin as a component of the beauty of the Japanese woman," while the Japanese women preferred "light-brown skinned men. . . . Many women distinguished between a beautiful man and an attractive man. A beautiful man is white-skinned and delicately featured like a kabuki actor. Although he is admired and appreciated almost aesthetically, he is, at the same time, considered somewhat "too feminine" for a woman to depend upon. . . . On the other hand, an attractive man is dusky-skinned, energetic, masculine, and dependable." In a questionnaire study of Caucasian college students in Wyoming, the men said that they prefered to date and mate women with light eyes and hair, and medium light skin, while the women preferred men with dark eyes and dark hair, and they disliked very light complexions in men.

Such studies find relative, not absolute preferences. In societies comprised of many racial groups, preferences for women with lighter skin color is useless per se as a fecundity detector. Racial differences will swamp any subtle differences among women due to age or fecundity or parity. And of course in multiracial societies a man's desire for a lighter-skinned woman may have little to do with his beauty detectors and much to do with his status aspirations or his racism. When black women in South Africa in the 1970s suffered an epidemic of a severe skin disorder called ochronosis because of their excessive skin bleaching, this was not caused by their overzealous attempts to look nubile, but by their desire to have lighter skin in a society where rights and privileges were tied to skin color.

Blushing Brides

Charles Darwin observed, "Blushing is the most peculiar and the most human of all expressions. Monkeys redden from passion, but it would require an overwhelming amount of evidence to make us believe that any animal could blush." Blushing, like ticklishness, is an attribute of the young that tends to disappear in maturity. Children and adolescents blush more than adults, and girls blush more than boys. The blush is confined to the very visible regions of the face, neck, and sometimes the chest. Science reporter Roger Bingham has suggested that blushing is an honest advertisement of nubility—a sign of self-conscious youth with sexual imagination and a sexual future but little sexual past. Although blushing can signal painful self-consciousness, it is a frequent companion to young love. As Darwin noted, "No happy pair of young lovers, valuing each other's admiration and love more than anything else in the world, probably ever courted each other without many a blush."

As V. S. Ramachandran points out, blushing, and flushing suggest sexual excitement. When coloring gets vivid, the skin is moist, the lips swell, and the skin generally signals "the likelihood that one's courtship gestures will be reciprocated and consummated." Red rather than pale lips and pink rather than white cheeks also advertise health. Anemia due to iron-poor blood is a common illness in most countries, and its telltale signal is pallor. Today many women become anemic after years of blood loss during menstruation, but in the environment of our ancestors that was a rarity. Women were either pregnant or lactating, and they seldom menstruated. But in those environments, anemia signaled either an iron-poor diet or a response to parasite infection (in infection, the body binds iron as an adaptive strategy).

Most women unthinkingly apply blush and lipstick every day, many add foundations and powders a shade lighter than their natural complexions. The light foundation and blush on the cheeks and red on the lips are sexual signals mimicking youth, nulliparity, the blush of youth, and the vigor of health.

Women often talk about "putting on" their face and are willing to cover their natural skin tone and texture daily with a small range of shades produced by cosmetics manufacturers. As anthropologist Marilyn Strathern notes, "The skin, the outer surface, is, in this context, truly superficial, trivial in relation to personal identity." When women put on face makeup they are reworking their faces to approximate a shared ideal, in fact to replace their individual feature (their skin's unique properties) with an idealized feature different from their own. Philosopher Stanley Cavell likens contemporary actresses to "face shapes that run through various hair styles on a barber shop chart . . . the new actresses tend to be cosmetics." The trade-off between revealing the uniqueness of the individual's appearance and "beauty," the willingness to give up some of the former to gain more of the latter, goes to further extremes as women leave the realm of temporary paints and powders for surgical alterations.

Skin Tight

Mimicking the fresh luminant skin of youth is not easy, but that hasn't stopped anyone from trying. A young person's skin looks fresh and it is: new skin cells push to the surface within two weeks. The skin is in perpetual bloom. As skin gets older, this process slows down. The surface cells, pushing at the boundary of their expiration date, look sallow and muddy. Oil glands become less active, and collagen and elastin break down, making the skin drier and less supple. Crinkles that once came and went with fleeting expressions etch themselves onto the face. The round cheeks of youth become thinner and more angular as the fat beneath the face thins. The skin sags.

If we are lucky enough to grow old, we will all experience skin aging. But people show remarkable differences in when the signs of aging occur. Fair-skinned Caucasians begin finding wrinkles about ten to twenty years earlier than African-Americans. Women tend to wrinkle before men do. But skin aging is not a simple unfolding of a genetic timetable. Health and habits play a large role. Nineteenth-century light-skinned belles hid their skins beneath parasols and cultivated their pallor. Twentieth-century beauties mocked these pale women as symbols of weakness and oppression. Starting in the 1920s, a tan became a sign of health and wealth, and of a life spent out in the world rather than within the shaded confines of home. Everyone would have been better off realizing that skin color is skin deep.

Nancy Burson and David Kramlich, an artist and a computer scientist, have designed what they call an "age machine," which takes an image of your face and uses computer technology to age it. I caught a glimpse of the machine when it came to the MIT List Art Gallery in 1990. It has two alternative visions of the future—one for people who take care of their skin, and one for people who sunbathe and smoke. Johnson and Johnson wisely took it to malls in 1990 to promote its new sunscreen and moisturizer. If anything could help sell the product, this might.

In the provocatively titled, "Does Cigarette Smoking Make You Old and Ugly?" Deborah Grady and Virginia Ernster reviewed the evidence in a 1995 article in the *American Journal of Epidemiology*. Their answer is yes. Heavy smokers develop more wrinkles than non-smokers, but not until they reach their forties. They are also more likely to turn prematurely gray, and men are more likely to go bald. Cigarette smoke is toxic to skin for many reasons. Nicotine constricts the blood vessels and blood flow to the skin, which can make the face as ashen as the tip of a cigarette. The frequent squinting, inhaling, and puffing that accompany cigarette smoking may etch permanent traces on the skin. Since cigarettes are known to damage collagen and elastin in the lungs, it is also possible that they directly damage these in the skin as well.

Tanned skin may look all aglow with health, but then so may feverish or burned skin. Tans and sunburns are signs of skin damage.

They occur in response to ultraviolet light, when the skin thickens and produces more melanin to protect itself. It is only later that the damage is visible in signs from wrinkles to brown spots to skin cancer. Only the natural amount of melanin in skin (the skin's color before tanning) can protect skin from damage, which is why African-Americans get skin cancer only about one-fourteenth as frequently as Caucasian non-Hispanics do.

Because the damages are not visible for years, many health-seeking, beauty-loving people continue to sunbathe and smoke. The rewards are immediate—nicotine makes a direct hit on the pleasure centers of the brain, and tans suggest the seaside adventures of youth and elicit compliments. Perhaps some of the highly visible people who smoke and sunbathe despite knowing the damages—such as actors and supermodels—are honestly advertising their health by purposeful handicapping. Like piercings and tattoos, their tans and cigarettes say, "I am so healthy I can do dangerous things and still be unlined, as yet unscathed, and beautiful." Twenty years later, the boast may catch up with them. But in youth it gives them an irresistible insouciance.

To counter the ravages of age and the residue of bad habits, women have hidden behind face makeup. But since prehistoric times attempts have been made to forestall aging and not just mask it. Perhaps it reflects the harshness of our age, but we are no longer content to soothe and moisturize and hydrate the skin. Today we are most likely to put acids from fruit or sugar cane (alpha hydroxy acids) on our faces to peel off the top layers of skin; or we use topical preparations made from vitamin A derivatives (such as Retin-A or Renova), which speed up the skin turnover. For those who want to erase ten years, acids give way to the knife or the laser beam.

Nearly half of the world's cosmetic surgeons are in America, and a third of those work in California. But plastic surgery is going mainstream. Seventy percent of cosmetic surgery patients earn less than fifty thousand dollars a year, and thirty percent earn below twenty-five thousand. More than half of the respondents to a 1993 *Health* magazine survey agreed that cosmetic surgery "would eventually be as routine as hair dyeing." After Helen Bransford heard her husband, writer Jay McInerny, mention a beautiful younger actress once too often, she got a face lift. She predicts that "by the year 2000 cosmetic surgery may well be viewed as no more than a technological extension of makeup." To put it in historical perspective, she says, "If anesthesia had been around, Cleopatra would have been the Cher of her millennium."

According to the American Society of Plastic and Reconstructive Surgeons over six hundred thousand cosmetic surgery procedures were performed in 1996. Most of these were done on Caucasian women in their thirties, forties, and fifties. People of color represent about twenty percent of cosmetic surgery patients. Men are as likely as women to get their protruding ears pinned back, and many men get their noses resculpted (twenty-four percent of nose jobs are given to men) but women swamp men when it comes to fixing up aging skin.

Seventy-four percent of the cosmetic procedures reported by the American Academy of Facial Plastic and Reconstructive Surgery in 1993 were done on women, as were eighty-nine percent of the cosmetic procedures reported by the American Society of Plastic and Reconstructive Surgeons in 1996. Although eye lifts and face lifts are among the five most popular procedures performed on men, eighty-five percent of eyelid lifts and ninety-one percent of face lifts are performed on women. As we have discussed, aging has a bigger impact on men's mate preferences. Most people believe that looks decline with age for both sexes, and studies tend to support this. But the decline is steeper for ratings of women than for men, and steepest of all when men rate photographs of women. It is no wonder that women go to such lengths to conceal their age.

Estrogen deficiency appears to be an important factor in skin aging, making women particularly sensitive to its signs. Loss of estrogen at menopause is commonly accompanied by loss of collagen, making the skin thinner, drier, and more prone to wrinkling. When gynecologist Rodolphe Maheux gave estrogen supplements to sixty nuns at the Good Shepherd Sisters convent in Quebec City, women with little sun exposure and no history of smoking, he found that the women increased the thickness of their skin by twelve percent within one year. Other studies have found that hormone replacement therapy decreases wrinkle depth and pore size and increases skin moisure and numbers of collagen fibers in postmenopausal women. No one is suggesting hormone replacement therapy solely as a beauty aid (especially since it may also cause skin breakouts and melasma) but the studies highlight the role of estrogen in keeping skin young and dewy.

In the future, we may lose a sense of what it looks like to age. Instead of letting faces age, people are now getting little "procedures" at the first signs of aging. The newest trend is called "age dropping." It means that people are starting plastic surgery in their thirties rather than in their fifties The idea is never to visibly age at all. Unlike Oscar Wilde's Dorian Gray, however, there will be no hidden painting in the attic charting the decline; the evidence will be tossed away (or harvested to use in future procedures).

Spotless

Looking at advertisements for pimple creams, and the waiting rooms of dermatologists' offices, one would presume that acne was a female problem. Women are much more likely than men to consult dermatologists about skin eruptions, and much more likely to use concealing products. However, it turns out that males tend to blemish more than females, and for a simple reason: the excess sebum (or oil) which leads to pimples is stimulated by androgens, the male hormones. Although androgens are called male hormones, they are produced by the adrenal glands in both sexes. Females produce small amounts in their ovaries, while males produce ten times as much in their testes.

All adolescents get pimples, which they hope to outgrow along with unpredictable growth spurts, cracking voices, and mood swings. But pimples may persist beyond the teenage years. In young adult women, acne may signal an abnormal increase in or sensitivity to circulating male hormones. In one study, ninety percent of the young women who sought treatment for acne had higher than average levels of testosterone, and more than half of them had ovulatory dysfunction. Many of the acne sufferers also had a tendency to develop excess hair on the face and other regions. In women approaching menopause, acne may reflect the fact that the estrogen level is too low to counter the effect of androgens.

If women blemish less than men do, they worry more over each spot. Perhaps blemishes are more alarming and are more carefully concealed because they are linked to the body's androgens, and anything linked to androgens (or masculinity) on women is a signal at cross-purposes with the advertisements of fertile beauty. This is not to imply that pimples are welcome by either sex: they are not. They are always signs of infection (even acne is caused by bacteria) and can signal more dire diseases (measles and so on) or skin parasites.

Blank Allure

The facial muscles lying right beneath the skin serve an important evolutionary purpose: they allow the face to show an exquisite range of feelings. This is why muscles on the face are attached directly to skin, unlike anywhere else on the body. But after years of repeated tugging and crinkling these feelings may start to linger longer than intended. Women's faces tend to be more animated then men's, and their skin more delicate, so that these expressions can take on an unwelcome permanence.

Many women complain that as they age they start to look not just tired but angry. One woman said that she was sick of hearing, "Smile, it's not that bad." Why does an older face look angry when the person doesn't feel angry? The lips get thinner as we age, and the brows lower. Both mimic what we do with our facial muscles when we get angry: we lower our brows and pull our lips into a tight line. When we are worried or straining, we may also bring our brows together and knit them. As we get older, the vertical lines between our brows and the horizontal lines across our foreheads leave a faint echo of feelings past.

One of the newer cosmetic surgery techniques involves injecting a small amount of the nerve poison that causes botulism into the forehead between the brows. This temporarily paralyzes the corrugator muscle, the one that knits the brows together. After the "botox" treatment, the vertical lines between the brows disappear, no matter how vexed or perplexed a person is. A more permanent procedure is to snip the corrugator muscle. But how does one accurately convey feelings after paralyzing facial muscles? The corrugator normally knits when we hear unpleasant sounds, see objects we abhor or are frightened by, or when we see another person's face with a knit corrugator. One way we communicate emotions is by facial mimicry, unconsciously adopting the expressions of others. Facial muscle activity is intimately related to our experience of emotion and is a readout of our reactions.

Botox injections and other paralytic procedures are not the only ones to tamper with facial expressions. Brow lifts can inadvertently give the face a look of permanent wide-eyed surprise. One study compared the actual placement of brows after brow lifts with the opinions of cosmetic surgeons and cosmetologists about the ideal placement of brows. When looking at computer-altered faces differing only in brow height and shape, both groups preferred eyebrows at or below the supraorbital rim and disliked eyebrows above the rim. One hundred postoperative photographs taken from sixteen frequently cited articles on brow lifts, however, show the eyebrow placed above the rim. This gives the woman a look of permanent surprise.

We will talk more about altering facial features in Chapter 5. Surgery on the skin tends to restore youth and erase signs of bad habits. But for the (mainly) women who opt for it, the change that may come from altering the upper third of the face is toward a more naive, open-eyed look and away from any sign of worry or anger. A lifted face looks perkier, but inevitably less commanding. The current vogue is for expressionless, blank-faced models. Diderot said that beauty in a woman is a vacant face, "a face of a young woman . . . innocent, naive, still without expression." Is this the face women want to put forward as they grow older?

Paul Ekman, the world's expert on facial expression, believes that the nuances of expression may be as much under genetic control as the shape of our noses or the curve of our lips. When we alter expression we may also be altering family resemblance, the signature displays that are as heritable as the sound of our voice or the way we laugh.

Beauty and Skin Color

Beauty and race are an explosive combination, a melding of two incendiary topics. But with white models peering out from fashion magazines all over the world, the question is inevitable: why? The version of *Elle* magazine distributed in China and Tibet (where it has eight subscribers) has Asian models inside but white models on the cover. A stranger familiarizing himself with Brazil through its magazines would, as a recent *New York Times* article said, "mistake this racial rainbow of a country for a Nordic outpost . . . slender blondes smile from the covers and white faces dominate all but sports glossies."

No one knows what the original color of skin was, but most people in the world today are brown, and it is likely that ten thousand years ago all people were brown. The amount of pigment or melanin in the skin helps protect it from sun damage. In equatorial Zaire, the dark-skinned natives rarely get skin cancer while in Australia, where descendants of Englishmen and Irishmen have transplanted themselves and their pale skins, skin cancer is more prevalent than anywhere else in the world.

Pale skin is an adaptation to low light, Since the body converts sunlight on skin to vitamin D, and then to calcium, very pale skin confers a survival advantage when sunlight is scarce by allowing the maximum amount of sunlight to penetrate skin. In parts of the world where the diet is rich in vitamin D, for example among the Inuit living in the Arctic who exist on a diet of fish and fish oils, the skin is not pale.

There is nothing inherently more beautiful about Caucasian skin and, in fact, it has many disadvantages compared to other skin types. It wrinkles earlier, has more freckles, and is more prone to acne and skin cancer than Asian or African skin. It is also more likely to be accompanied by head hair that grays earlier and more extensively than African or Asian hair. Caucasian men are the most likely to go bald, and Caucasians have more body hair than any other racial group, except the Ainus, an ancient tribe in northern Japan.

Many scientists such as Jared Diamond believe that geographical differences in skin tones and other physical features might reflect the workings not only of natural selection (as adaptations to climate and environment) but of sexual selection. Our ancestors developed in isolation from one another, and their local, arbitrary preferences may have spawned future generations who varied more dramatically in their features than may have occurred if features varied solely in response to climatic conditions.

Whether or not sexual selection played a key role in the range of skin tones we see today, one thing is clear: they reflect little about what goes on under the skin. As geneticist Luigi Luca Cavalli Sforza writes, "It is because they are external that these racial differences strike us so forcibly and we automatically assume that differences of similar magnitude exist below the surface in the rest of our genetic makeup. This is simply not so: the remainder of our genetic makeup hardly differs at all." But anthropologist Alan Goodman is quick to make the distinction between the facts of science and the prejudices of the human mind: "Race as bad biology has nothing to do with race as lived experience. . . . True races may not exist but racism does."

Beauty judgments are sensitive barometers of social status. In all countries the economically dominant group has put forward its own ethnic features as the standard of beauty, and in widespread dominance mimicry, other groups tend to follow the group's lead. The universal preferences remain—for clear skin, lustrous hair, full lips, and so on—but the exact incarnation of these features can differ depending on who holds the reins of power. When studying race relations in the West Indies in the 1960s, sociologist Harry Hoetink observed that the standards of physical attractiveness were always molded by the appearance of the dominant group. Those who can "pass" as members of the group in power are more likely to rise in status and be considered "attractive" by that society's standards.

Why do light-skinned models adorn magazines in Brazil? As in the United States, the phenomenon can be traced to the radical inequality between the Portuguese, who arrived in 1500, the indigenous Indian population, whom they conquered, and the Africans, many of whom were brought as slaves for the sugar plantations. Four centuries later, only forty percent of Brazil's citizens are white but they remain the rich and powerful. In 1996 a magazine called *Brazil Race* was launched for the "invisible 90 million people"—the nonwhites who are seldom seen in the media. It sold out its first run of 200,000 copies within a week, prompting its editor, Aroldo Macedo, to say that it single-handedly exploded the myth that a magazine with blacks on the cover would never sell.

In the United States the standard of beauty has been not only white, but white as exemplified by the first wave of immigrants, the northern and western Europeans. European settlers decimated the native Indian population and brought Africans as slaves, setting up a power elite based on race. Later immigrants in the early twentieth century often arrived from southern and eastern European countries such as Italy, Poland, and Russia. Current immigration is highest from Asia, Central and South America, and Africa. American beauties have reflected the fact that northern and western Europeans arrived first and established themselves as the elite. The ideal of beauty has been not only white but WASP. In 1921, the first Miss America was Margaret Gorman, a fifteen-year-old blue-eyed, blond high schooler. The first and only Jewish Miss America was Bess Myerson in 1945. It wasn't until 1984 that the crown went to an African-American woman, Vanessa Williams, albeit one who had light skin and hazel eyes.

Cultural commentator M. G. Lord has called the Barbie doll "a space age fertility icon" who "has come to represent not merely American women or consumer capitalist women but a female principle that defies national, ethnic, and regional boundaries." However, from 1959 until 1980 that female principle was blond and blue-eyed. It took twenty years for Mattel to launch Black Barbie and Hispanic Barbie. Coincidentally, they did it in the same year that the first African-American woman became a *Playboy* Playmate of the year.

Asked in 1969 if African-American models were a "trend," New York modeling agency head Wilhelmina, who represented one of the first black models to don a magazine cover, Naomi Sims, said angrily, "No, because Negroes aren't temporary." But over twenty years later women of color were still finding few models whose facial features mirrored their own. A 1991 report entitled, "Invisible People," published by the Department of Consumer Affairs in New York City, surveyed over eleven thousand advertisements in twenty-seven different national magazines, and one hundred and fifty-seven fashion catalogues. They found that ninety-six percent of the models were white. Although eleven percent of the magazines' readership was African-American, three percent of the models in advertisements and five percent of the models in editorials were African-Americans and only one percent of the models were Asian.

In 1994 the Women in Media group, founded by Betty Friedan and Nancy Woodhull to study gender issues in media, surveyed the photographs in nine women's magazines After looking at the October issues of *Allure, Cosmopolitan, Elle, Essence, Glamour, Harper's Bazaar, Mirabella, Vogue,* and *Ladies' Home Journal,* they had had their fill of skinny white women. They'd seen two hundred and five of them. They'd also seen but ten African-American models and six "Asian and other" women.

Europeans and their descendants are unlikely to maintain their dominance forever. In the United States they will soon be outnumbered and perhaps outspent. A recent *Newsweek* article predicted that by the year 2050 the percentage of white, non-Hispanic Americans will plummet from seventy-four percent (the percentage in 1995) to fifty-three percent. There will also be increased intermarriage. In 1990, sixty-seven percent of Asian-Americans in their twenties who were born in the United States married outside their race. In 1996, African-Americans' total earned income was three hundred and sixty-seven million dollars and they spent three times more per capita than other consumer groups on cosmetics, toiletries, and grooming. African-Americans purchased thirty-four percent of all hair care products sold in the United States in 1992. But even though African-Americans spent one hundred and seventy-five million dollars on magazines, advertisers still claim that a black woman on the cover will decrease sales, and magazines such as *Ebony Jet* and *Ebony Man* which aim at the African-American audience have yet to attract advertisements from the high-end fashion business.

As anthropologist Douglas Jones writes, "As long as societies are stratified and physical features covary [correlate] with social status, somatic norms will have an effect on images of beauty. The individuals' and groups' social positions—especially their bargaining power in the mating and marriage market—may depend not just on their economic and political assets but on their somatic distance from dominant groups." It is likely that All-American beauty will look different as we head into the next century and that the iron grip of the northern and western European blonde will give way to greater diversity. The definition of a beautiful cover girl as a white girl may go the way of the Crayola "flesh tone" crayon.

Hair

A full head of human hair contains about one hundred thousand strands, each growing six inches a year and eventually reaching two to three feet in length. Head hair gives us padding and protection, but why do we need three feet of it? Eyelashes are a modest fringe that keep out the glare and grit from our eyes. Eyebrows shield the eyes from sweat and sun, and may have evolved in part for their communicative function in facial expressions. Body hair is a signal of sexual maturity. Head hair doesn't have much use at all—except in attracting mates.

When a woman starts playing with her hair, flinging and flipping it, she is signaling sexual interest. Social scientists note that girls on the prowl always show off a trio of come hithers—they lick their lips, toss their heads, and flip their hair. Hair has the whole sensory package working for it: it has color, shine, texture, perfume, and movement.

In 1548, Agnolo Firenzuola, an Italian monk and author of *Dialogo delle Bellezze delle Donne (Dialogue on the Beauty of Women)*, wrote, "However well-favored a lady may be if she have not fine hair, her beauty is despoiled of all charm and glory." Although flawless skin may be the most universally desired feature, hair runs a close second. Over half the women who answered a 1993 *Glamour* magazine survey agreed with the statements, "If my hair looks good, I feel attractive no matter what I'm wearing or how I look otherwise," and, "If my hair isn't right, nothing else can make me feel that I look good."

Women's hair is considered so sexually provocative that in many cultures it is concealed after marriage for fear of inciting uncontrollable desires. In the first century a married Roman woman could be divorced for uncovering her head. The Talmud says that a woman may be divorced without payment of marriage contract if she goes out with her hair fully or partially uncovered. To this day, Orthodox Jewish women, once married, cover their hair with a kerchief or scarf or wear a wig. Nuns, married to Christ, cover their hair and hide it in the coif. Greek women in the Doric period cut their hair on the day of their marriage. In sixteenth-century Italy married women wore their hair loose and free until their marriage, at which point they put on veils and scarves or caged their hair in nets and coils.

Men also have a complex relationship with their hair. Samson, who lost all power when shorn of his locks, casts a long shadow. In 1990, psychologist Thomas Cash studied people's perceptions of men who had either a full head of hair or were balding. He found that both sexes assumed the balding men were weaker and found them less attractive. No wonder the same author found in a different study that seventy-five percent of the bald men he interviewed felt self-conscious about their baldness, and forty percent would wear a hat to hide it. Napoleon's valet related that, when Napoleon met with Tsar Alexander of Russia to discuss European politics, the two ended up talking about cures for baldness.

Hair Raising

A lot of time and effort have gone into exaggerating the already natural abundance of human hair. Big hair is a competitive enterprise from the highlands of New Guinea to the shopping malls of the United States. In the highlands of New Guinea tribesmen think that the ghosts of ancestors lodge in the hair and that baldness is a sign that the ancestors have abandoned a man. When they court women, they build large wigs made of hair mixed with clay and then sewn onto a frame of cane, hardened with dipped wax, painted and adorned with vines, beetles, side ringlets, and fur.

The term "bigwig" came about because of the habit of the European male aristocracy of wearing enormous fake hair. In the late seventeenth century and into the eighteenth century men's wigs were parted in the middle, with twin peaks on either side of the part, and curls cascading to the shoulder and beyond. A satirist described the appearance of the face as but "a small pimple in the midst of a vast sea of hair."

When John Travolta gets himself ready for the disco floor in *Saturday Night Fever*, he warns his father, "Don't touch the hair." Rock and roll heroes don't seem to get anywhere without eye-catching do's. James Brown explained that when people ask why he wears his hair so high: "I tell them so people won't say where he is, but THERE he is." The Beatles were first made recognizable by their mop tops. Long hair later signaled male rebellion, the counterculture, and Samson-like power.

In John Waters's hilarious send-up of sixties hairstyles, *Hairspray*, Debbie Harry's hair is so big, she can conceal a bomb beneath it. Over the course of history women have extended their hair in every possible direction. When the Infanta Maria Theresa arrived in France to marry Louis XIV in the mid-seventeenth century, she wore her hair low on top, medium in length but extremely wide at the side, a style that can be seen in the paintings of Diego Velasquez. Three centuries later, some American bouffants in the mid-1950s were said to be as much as fourteen inches wide. The wide do is less popular than long flowing tresses or piling the hair skyward. The tallest hair was created by European aristocrats in the late eighteenth century. Hair had become a work of art among aristocratic women, and it was stuffed with wool or horsehair pads or wires and kept in place by pomade and flour, and decorated with tiny ornaments depicting landscapes and battle scenes. In 1780 the doorway of St. Paul's Cathedral had to be raised four feet to accommodate big-haired women. Women had to crouch in carriages because their heads were too big to sit, and they had to sleep on their backs in order not to ruin the do.

Let Your Hair Down

Hair on our head grows about half an inch a month. It grows fastest in young adults, and fastest of all in girls between the ages of sixteen and twenty-four, and slows down in middle age. If left uncut, hair naturally grows to about two or three feet before falling out. According to a 1949 report in the *Toronto Star*, the longest hair on record belonged to an Indian monk named Swami Pandarassannadhi with twenty-six feet of hair—about the length of hair that a fifty-year-old man would have if he never cut it and it never fell out.

Most men prefer long hair on women, and many of the male romantic heroes such as Fabio, Yanni, the recently shorn Michael Bolton, romance novel heroes, and so on, have flowing tresses. As the Parisian hairdresser Croisat said in the early nineteenth century, "You never see Eve, Venus, or the Graces painted with cropped hair." Milton portrayed Eve in *Paradise Lost* with gold hair that fell in ringlets. Botticelli and Titian nudes have wild flowing hair. In ancient mythology, Lorelei sang on a rock in the Rhine, and the liquid gold of her hair lured boatmen to their death if they but raised their eyes to look at her.

One of the reasons we love long hair is that it is one of the most informative aspects of a person. Certainly hair reflects a lot about our attitudes, affiliations, self-esteem, taste, gender, age, and so on. But it is also a living record of our bodies over a period of time. Your outer layer of skin may slough off within a month, but if your hair falls past your shoulders it has been on your head for a couple of years. Unless we crop our hair or shave our heads, it's giving signs of what we have eaten, what drugs we have taken, and provides a record of how our health has been for as long as it grows on our head. It's a bit disconcerting, but part of the beauty of long hair may be that it has a history, our history, written on it.

And the longer it grows, the more it reveals. One hundred and sixty-six years after John Keats died, Dr. Werner Baumgartner analyzed a lock of his hair and found traces of the opiate laudanum. Each strand of hair has its own blood supply and reflects whatever is coursing through us, which is why drugs from aspirins to anticoagulants to thyroid medications can affect the health of hair, and why drug-testing companies are now experimenting with hair analysis rather than urinalysis.

Fleeting Fleece

We don't think of women as going bald, and most don't. But everyone's hair thins with age, both the diameter of each individual hair and the overall number of hairs. Although men go bald more prominently, most women have some loss of hair as well. Stress is one culprit, causing an increase in adrenaline that may lead to production of cholesterol and the male hormone testosterone. Hair loss becomes more marked in women after menopause.

At all ages, hair is influenced by health and diet. In a poor grazing season an animal has thinner fleece. Among merino sheep, fleece weight can vary up to four hundred percent between a good and a poor season. The potential is set by genetics—no matter how well you eat, you cannot exceed your genetic maximum, but there is room for tremendous variation beneath the ideal. Anorexics often notice a loss of head hair from starvation. Lack of copper, zinc, iron, vitamins A and E, and other nutrients will thin and damage the hair. In disease, a body may need to sequester nutrients. Since hair is nonessential (when compared, say, to bone marrow), a shortage of iron or of protein will be reflected in the hair. Hair is under the delicate control of the hormones, which is why men have beards and hair on their chests and male-pattern baldness on their heads, and women usually don't. Paradoxically, hairless heads and hirsute bodies can both be traced to secretion of androgens; eunuchs castrated before puberty never go bald. Little comfort that it is to a bald man.

No matter how high a level of androgen a person has, he won't go bald unless he is genetically predisposed to do so. About one fifth of men are balding by age thirty but another fifth still have a full head of hair after age sixty. The rest show a gradual thinning. Although there is no magic bullet to cure baldness, it's an area of intense scrutiny. Minoxidil was approved as a treatment for hair loss by the FDA in 1988. It was a case of turning a sow's ear into a silk purse. When used as an oral treatment for high blood pressure, it caused hair to grow on the forehead and across the bridge of the nose. About twenty-five percent of users who apply Rogaine (which contains minoxidil) to their heads grow back a moderate amount of hair (which falls out if they discontinue Rogaine). A new medication called Propecia, which acts directly on the hormones, is being touted as the first drug to stop hair from falling out.

The makers of Rogaine became concerned about reports of heart attacks among minoxidil users and sponsored a study of over six hundred men under the age of fifty-five who were admitted to hospitals in Massachusetts and Rhode Island for a nonfatal heart attack. They discovered that certain forms of baldness are associated with increased risk for heart disease (independent of minoxidil use). Although frontal balding may be the most aesthetically disturbing to the man, severe balding at the vertex (at the top of the head) was the only pattern associated with cardiac risk. The authors of the study speculate that a common cause for both the baldness and the risk of heart attack may be the male hormone dihydrotestosterone. Before covering up the spot, men should heed its warning.

So far, no one has come up with a cure for baldness. In 4000 B.C. the remedy was to rub the head vigorously with dog paws, dates, and asses' hoofs that had been ground up and cooked in oil. Today men apply Rogaine or get hair weaves or "hair replacement systems." They may do a comb-over or a comb-forward or some other technique to cover bald patches. There are forty million bald men in the United States, and not surprisingly, hair transplants are the most common plastic surgery for men.

One current trend for men is shaving or cropping the hair very close. The basketball star Michael Jordan has a shaved head as do many other athletes and some actors. It obscures where the hairline really is and takes away all evidence of the hair itself (whether it was thin or thick, gray or colored). Instead of bald and hairy patches, there is one smooth surface. One problem with male-pattern baldness is that the hair often recedes from the front. The receding hairline is a constant reminder that hair once was there, and that it is fleeing. We make an unconscious attempt to see the face as it was when it was framed by a full head of hair. By taking away the hair, we now radically adjust to the head as a whole. Does anyone remember how Michael Jordan looked with hair? Or Yul Brynner?

Shaving the head is a preemptive strike, a bold move to erase a sign of aging. The absence of hair on the head also serves to exaggerate signals of strength. The smaller the head, the bigger the look of the neck and the body. Body builders often shave or crop their hair, the size contrast between the head and neck and shoulders emphasizing the massiveness of the chest.

Blonde Crazy

Across all hair-coloring manufacturers there are five hundred different shades of blond alone, from the strawberry blond to the platinum blond and every shade in between. These companies estimate that in the United States up to forty percent of women add blond to their hair. Sociologist Grant McCracken has written that "this culture has turned blondeness into a beacon and wired it into the navigational equipment of every male." Blond is a rare hair color naturally, though much more common in children. The average blonde has more individual hairs and finer hairs than a brunette or a redhead, about 140,000 hairs compared to a brunette's 108,000 or a redhead's 90,000.

Milton's Eve and Dante's Beatrice were both blondes. Ancient Egyptians in Africa wore wigs of gold. Greek and Roman women developed a passion for the fair hair they saw on the northern Gauls and got wigs made from it. Golden hair became the hair of the aristocrat and at other times the symbol of the prostitute. Fairy tale heroines, brides on wedding cakes, and images of the Virgin are all very likely to be blond.

Blondness exploded when the technology became available to bring it to the masses. The first widely available commercial dyes came to the market in 1930. Jean Harlow in 1931 put platinum blond on Hollywood's map. Hollywood then featured a string of movies about blond hair such as *Blonde Venus, Blonde Crazy, Blonde Fever, Platinum Blonde,* and *Blonde Trouble.* In 1953 the most famous blond movie of all appeared: *Gentlemen Prefer Blondes.* Blond dyes became available for home use by the 1950s. Companies went all out in exploiting blonde appeal, with campaign slogans such as: "If I have one life to live let me live it as a blonde." "Blondes have more fun." Clairol made famous the coy line, "Does she or doesn't she?" invoking the woman's hair color and her sexual availability in one teasing come-on.

V. S. Ramanchandran says that blond hair just happens to accompany light skin and, since men favor light skin, they favor blond women. Others think there is something about the hair itself. Like pale skin, it evokes the innocent and the young. Until Hollywood made blond hair sexy and dangerous, blond heroines tended to be sweetness and light. It was the dark-haired woman who was knowing and dangerous. In the Archie comics, sweet Betty is blond while shrewd Veronica is dark-haired. Scarlett O'Hara is raven-haired, while Melanie is a bland blonde in *Gone With the Wind.* In Disney's *Cinderella,* the good Cinderella is blond and blue-eyed, unlike her evil stepsisters and wicked stepmother, who are dark or red-haired. Rapunzel, Goldilocks, and a string of fairy tale heroines have blond hair. The only pure one who is a brunette is Snow White, whose mother's wish was to have a child "as white as snow, as red as blood, and as black as the wood of the window-frame."

When people are asked to rate various personality traits of people they see only in pictures, they tend to judge blondes as weaker, more submissive, and less wise. Is this a result of media stereotyping of the typical blond heroine? In an intriguing set of studies done on temperamental differences in infants and young children, psychologist Jerome Kagan has found that children with pale pigment, in particular children with blue eyes, are far more likely to be shy and inhibited than dark-eyed children. They are the most likely to be fearful of new situations, hesitant in approaching someone, quiet with a new person, and the most likely to stay close to their mothers. Brown-eyed children are bolder. Kagan speculates that fear of novelty, melanin production, and corticosteroid levels share some of the same genes.

His theory is speculative, suggesting that when people migrated to northern Europe they were faced with the problem of keeping up a body temperature that was used to a warmer climate. A mutation that increased the efficiency of the sympathetic nervous system and upped the level of norepinephrine (one of the major neurotransmitters) would have also raised their body temperature and offered a survival advantage. Unfortunately, it would also have left them with a more reactive nervous system and a more timorous temperament. Where does the pigment come in? High levels of norepinephrine can inhibit the production of melanin in the iris and can increase the level of circulating glucosteroids that can inhibit melanin production as well. So blond hair and blue eyes and shyness may be a common biological package. This may help explain the purity and innocence of the standard image of the blonde. Whether it explains the appeal of blondes to men, we can only speculate.

Good Hair, Bad Hair

When African-American models such as Naomi Campbell or entertainers such as Tina Turner wear blond hair, some people criticize them for trying to look like Caucasians in order to succeed, or accuse them of having internalized a Caucasian standard of beauty. Drag queen RuPaul waves the criticism away: "When I put on a blond wig, I am not selling out my blackness. Wearing a blond wig is not going to make me white. I'm not going to pass as white, and I am not trying to. The truth about the blond wig is so simple. It really pops. I want to create outrageous sensation, and blond hair against brown skin is a gorgeous outrageous combination." Basketball star Dennis Rodman is not trying to look like a white person any more than his former lover Madonna is trying to erase her Italian ethnic identity when they each bleached their dark hair blond.

But hair is a politicized issue for African-American women, and the decision whether to leave hair natural and wear it in an Afro, or dreadlocks or cornrows, or to straighten it is the source of much debate. Until the 1960s, most African-American men and women straightened their hair, and seventy-five percent of African-American women still process their hair with straightening combs and chemical relaxers.

Only recently have African-American women had a choice of natural styles if they wanted long hair. Cornrows and dreadlocks were not commonly seen. Now they are so accepted that they've crossed over and white blond women cornrow their hair, and young white guys with trust funds wear their hair in dreadlocks; in the ski towns of Colorado where they spend their time, they are known as trustafarians.

Among African-American women with long hair, it is upper-middle-class blacks who wear dreadlocks, twists, and Afros, according to psychologist Shanette Harris. This suggests that, with increased economic power, women will be more likely to wear natural hair, and that at least some proportion of hair straightening is for purposes of conformity and hopes of rising in the middle class. The great hair debate rages in the African-American community, and issues of hair acceptance are much equivalent to weight acceptance issues within the Caucasian community. Hair is loaded with meaning.

CHAPTER FIVE

Feature Presentation

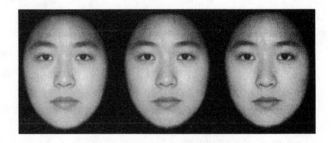

I'm happy people find me attractive, but really it's a matter of mathematics the number of millimeters between the eyes and chin.

— PAULINA PORIZKOVA

The human face is really like one of those Oriental gods: a whole group of faces juxtaposed on different planes; it is impossible to see them all simultaneously.

— MARCEL PROUST

In 1574, Hortensia Borromeo received a portrait from her husband, who was traveling far from home. In response, she wrote, "I experienced so much sweet emotion when I saw this picture. . . . Marveling, and repeatedly looking at the beautiful face, I . . . lost recollection of all else in the world." Queen Victoria was known to wear a signet ring with five tiny photographs of the faces of her family magnified by a jewel lens. Nothing captures our attention like a human face, and nothing rivals the face in communicative power.

We inherit our facial features and wear the noses, eye shapes, and chin angles of our ancestors. Anthropologist Melvin Konner suggests that the original adaptive basis of our interest in faces is to unravel kinship. He calls it the "are you related to so-and-so effect." But our borrowed features assemble into novel configurations, and each has its own stamp. Age, health, and habits etch in plot lines, and momentary pleasures and terrors ripple its surface. We become experts at ferreting out the curves and angles that distinguish one face from another and at gauging the quarter-of-an-inch movement of the brow or lip that signals a change from one mood to the next. As scientist Francis Galton said in 1883, "The difference in human features must be reckoned great, inasmuch as they enable us to distinguish a single known face among those thousands of strangers, though they are mostly too minute for measurement. The general expression of a face is the sum of a multitude of small details, which are viewed in such rapid succession that we seem to perceive them all at a single glance."

We notice beauty in a face quickly and automatically. Donald Giddon, a lecturer at the Harvard School of Dental Medicine, created a computer program that displays a face in profile and allows the user to modify its parts. When the user starts the computer animation, a receding jaw may begin to move forward. The user clicks the computer mouse to indicate when the feature looks acceptable, clicks when it is at its "most pleasing point," and releases the mouse when it is unacceptable. Giddon found that perceptions of facial attractiveness changed drastically with very minor alterations to features. Moving a feature one twenty-fifth of an inch (one millimeter) could flick the brain's switch from pleasing to unacceptable.

The features that signify our identity and ancestry evolved partly as adaptations to climatic conditions, just as body shapes and skin tones did. Noses carry air into the lungs. They evolved into long narrow shapes in climates where the air was cold or dry and needed to be warmed and moistened before reaching the lungs. People of northern European or Middle Eastern ancestry often inherit long noses with narrow nostrils (perfect for restricting air flow). In humid environments, the short wide nose common to many African and Asian people are more efficient.

Eyes need extra protection in cold sunny climates where glare reflects off snow. Narrow eyes with an extra padding of fat surrounding them have built-in sunshades and are common among people of northern Asian ancestry. Other aspects of appearance that vary with race or geographic location, such as eye color, seem to have no biological function and probably reflect the arbitrary mating preferences of our ancestors or historical accidents. For example, in small isolated communities, the traits of the few individuals who founded the community dominate later generations: biologists call this the "founder effect."

When Europeans came into contact with people in Asia, Africa, and the South Pacific, all were stunned by the differences in facial morphology, and their reactions were generally unfavorable. Anthropologist Edward Westermarck describes one Tahitian as saying, "What a pity it is that English mothers pull their nose so much and make them so frightfully long." Darwin was told by his colleagues that Chinese people found the "prominent" Western nose "hideous." Meanwhile, the Europeans found the Chinese noses "very broad," and Darwin's colleagues reported that the Hottentots, Malays, Brazilians, and Tahitians "compress the noses and foreheads of their children for the sake of beauty."

Eyes were equally controversial. When Japan opened its doors after over two hundred and fifty years of isolation, the initial delegation of samurai sent to the United States in 1860 reported that Western women's eyes looked like "dogs' eyes" and that this was "disheartening." Westerners meanwhile were dismayed by the fact that half of the Asians lacked an upper eyelid crease, and some had an epicanthal fold, a piece of skin that partially covers the inner corner of the eye. This made Asian eyes look expressionless to Westerners, who were accustomed to eyelids that changed shape with different emotions. The narrower shape also made the eyes appear sleepy and small.

Skin color was perhaps the most salient difference. In many parts of the non-Caucasian world, Europeans seen for the first time were thought to be ghosts or ancestors returned from the dead. When Australian Michael Leahy explored the mountainous regions of New Guinea in 1930 he came across tribesmen who were "utterly thunderstuck by our appearance. . . . One old chap came forward gingerly with open mouth, and touched me to see if I was real. Then he knelt down, and rubbed his hands over my bare legs, possibly to find if they were painted." Unconvinced that these wan intruders were human, they spied on them to see if they defecated. When they did, the scout returned with the message, "Their skin may be different, but their shit smells bad like ours." Darwin reports, "The African Moors . . . knitted their brows and seemed to shudder" when they saw white skin because, here too people believed that white skin was only found in demons or spirits.

The Europeans and their American descendants in the eighteenth and early nineteenth centuries carried strong presumptions about the beauty and superiority of their race. Indeed, Darwin found it necessary to tell his readers that "savages" admire the beauty of their women: "I have heard it maintained that savages are quite indifferent about the beauty of their women. . . . This conclusion does not at all agree with the care which the women take in ornamenting themselves, or with their vanity." He also explained that the people of each culture preferred their appearance to the appearance of the Europeans. As one source told him, he "did not think it probable that Negroes would ever prefer the most beautiful European woman on the mere grounds of physical admiration to a good-looking Negress." Another told him that in Thailand the women "have small noses with divergent nostrils, a wide mouth, rather thick lips, a remarkably large face. . . . Yet they consider their own females to be much more beautiful than those of Europe."

There are many reasons why people may favor their own racial features. One is racism. Another is that features that are typical in one race may signal illness or deformity in another. White skin in an African or epicanthal folds in a Caucasian are extremely rare and are associated with genetic conditions such as albinism. Features typical of one race may simply represent extremes in normally encountered sizes or shapes and thus appear freakish. If we are used to seeing faces with short wide noses, long thin noses will capture our attention because of their rarity. This is one reason why people from other races "all look alike" when we encounter them for the first time—not because they look different, but because they all look different in the same way. London's National Portrait gallery contains a room filled with members of the eighteenth-century Kitkat club all wearing the same spectacular white wigs. It's impossible to tell one from another. But stand in that room long enough, and individuals rather than Whigs in wigs stare back. The eye calibrates and begins making its usual distinctions. It is probably not unlike an experience with a face of an unfamiliar race at first contact.

Most people find it easier to recognize members of their race than members of other races, a well-verified finding in research on eyewitness testimony. We tend to make "false positive" identifications of people of other races, thinking someone looks familiar who is not. We even have slightly greater difficulty distinguishing men's faces from women's faces in other races. This does not appear to be a product of our racial attitudes, but of our level of contact with members of other races. It arises out of learning, the process of becoming an expert at recognizing faces whose features and hair colors and textures vary in limited ways.

Universal Beauty

Despite racism, misperceptions, and misunderstandings, people have always been attracted to people of other races. Today, the world is a global community where international beauty competitions have enormous followings (although many complain that these contests favor Western ideals of beauty). There must be some general understanding of beauty, however vaguely defined, since even three-month-old infants prefer to gaze at faces that adults find attractive, including faces of people from races they had not been previously exposed to. In recent years scientists have taken a deep interest in the universality of beauty.

It turns out that people in the same culture agree strongly about who is beautiful and who is not. In 1960 a London newspaper published pictures of twelve young women's faces and asked its readers to rate their prettiness. There were over four thousand responses from all over Britain, from people of all social classes and from ages eight to eighty. This diverse group sent in remarkably consistent ratings. A similar study done five years later in the United States had ten thousand respondents who also showed a great deal of agreement in their ratings. The same result has emerged under more controlled conditions in psychologists' laboratories. People firmly believe that beauty is in the eye of the beholder, and then they jot down very similar judgments.

Our age and sex have little influence on our beauty judgments. As we have seen, three-month-old babies gaze longer at faces that adults find attractive. Seven-year-olds, twelve-year-olds, seventeen-year-olds, and adults do not differ significantly in their ratings of the attractiveness of the faces of children and adults. Women agree with men about which women are beautiful. Although men think they cannot judge another man's beauty, they agree among themselves and with women about which men are the handsomest.

Although the high level of agreement within cultures may simply reflect the success of Western media in disseminating particular ideals of beauty, cross-cultural research suggests that shared ideals of beauty are not dependent on media images. Perhaps the most far-reaching study of the influence of race and culture on judgments of beauty was conducted by anthropologists Douglas Jones and Kim Hill, who visited two relatively isolated tribes, the Hiwi Indians of Venezuela and the Ache Indians of Paraguay, as well as people in three Western cultures. The Ache and the Hiwi lived as hunters and gatherers until the 1960s and have met only a few Western missionaries and anthropologists. Neither tribe watches television, and they do not have contact with each other: the two cultures have been developing independently for thousands of years. Jones and Hill found that all five cultures had easily tapped local beauty standards. A Hiwi tribesman was as likely to agree with another tribesman about beauty as one American college student was with another. Whatever process leads to consensus within a culture does not depend on dissemination of media images.

Cross-cultural studies have been done with people in Australia, Austria, England, China, India, Japan, Korea, Scotland, and the United States. All show that there is significant agreement among people of different races and different cultures about which faces they consider beautiful, although agreement is stronger for faces of the same race as the perceiver.

In the Jones and Hill study, people in Brazil, the United States, and Russia, as well as the Hiwi and Ache Indians, were presented a multiracial, multicultural set of faces (Indian, African-American, Asian-American, Caucasians, mixed-race Brazilian, and others).

There was significant agreement among the five cultures in their beauty ratings and some differences. For example, the Hiwi and the Ache agreed more with each other than they did with people in the Western cultures. This is not because they share a culture—they don't—but because they have similar facial features, and they are sensitive to the degree of similarity between their facial features and the features of the people in the photographs. For example, although the Ache had never met an Asian person, they were curious about the Asian-American faces, attracted to them, and aware of the similarity between these faces and their own. The Ache gave less favorable ratings overall to African-American faces, and they called the Caucasian anthropologists "pyta puku," meaning longnose, behind their backs. One Caucasian anthropologist was given the nickname "anteater."

Since the Hiwi and the Ache had never encountered Asians or Africans, had met only a few Caucasians, and were not accustomed to using scientists' rating scales, any level of agreement with the Western cultures is intriguing. Jones found a number of points of agreement. People in all five cultures were attracted to similar geometric proportions in the face They liked female faces with small lower faces (delicate jaws and relatively small chins) and eyes that were large in relation to the length of the face. Jones calls these "exaggerated markers of youthfulness," and they are similar to the features mentioned in other cross-cultural studies of beauty. For example, psychologist Michael Cunningham found that beautiful Asian, Hispanic, Afro-Caribbean, and Caucasian women had large, widely spaced eyes, high cheekbones, small chins, and full lips.

People tend to agree about which faces are beautiful, and to find similar features attractive across ethnically diverse faces. The role of individual taste is far more insignificant than folk wisdom would have us believe. Although evolutionary psychology has not yet been able to determine the exact face of beauty, the research we are about to review suggests that as anthropologist Donald Symons puts it, beauty may be in the adaptations of the beholder.

Beauty by the Numbers

For thousands of years the answer as to what constitutes beauty would have been: numbers. As we have seen, mathematical ideals of beauty stretch back to Pythagoras and Plato, and to Dürer, da Vinci, and other artists of the Renaissance. At the heart of the classical notion of beauty was unity and order. In Vitruvius's *De architectura*, he describes what he calls "the well-shaped man," and gives him a face that divides evenly into thirds, and a head height that is one eighth of his total height. In his sixteenth-century treatise, *De divina proportione*, Franciscan friar Luca Pacioli proposed that the human body contained in microcosm the formula for the beauty of all things: "from the human body derive all measures and their denominations and in it is to be found all and every ratio and proportion by which God reveals the innermost secrets of nature." This book was illustrated by Leonardo da Vinci and contains his famous drawing of Vitruvian man as ideal form, with his arms and legs extended, the body fitted into a perfect square and circle.

Cosmetic surgeons are as likely to have seen Da Vinci and Dürer's formulations on beauty as to have perfected sutures. They were the ones who eventually put these ideas to the test when they undertook the daunting task of reconstructing faces and making them more beautiful. What was to guide them? As one plastic surgeon has said, "I'd so often plan out a facial reconstruction . . . and then agonize for days, not knowing for sure whether they would be better looking or not."

But as we saw in Chapter 1, the Renaissance canons proved in anthropometrist Leslie Farkas's words "not entirely realistic." In his exhaustive study of nine canons, Farkas measured hundreds of women's faces to see if they divided into equal thirds or fourths at particular landmarks. He calculated the relations among the facial features to see if the width of the nose was equal to the distance between the eyes, and the distance between the eyes was equal to the length of each eye. He tested whether the mouth was one and a half times as wide as the nose, and the nose one quarter the width of the face. The canons also predicted that the heights and inclinations of the ear and nose would be equal.

Some of these proportions were not found on any person, or rarely found. Some of the proportions were not pleasing, and others did not distinguish between the faces of women who were attractive or unattractive. Farkas measured only Caucasian women's faces. When cosmetic surgeon W. Earle Matory, Jr., measured the faces of four hundred attractive people from many different racial and ethnic groups, he found that only a few Caucasians with narrow features fit the classical canon for nose shape (nose width equal to the distance between the eyes). Noses this narrow were virtually never seen in attractive Asian, African-American, or Hispanic faces. The noses of Asians and Africans are usually wider and have a broader tip than the typical Caucasian nose and they project at a slightly different angle. Cosmetic surgeons used to go so far in their effort to narrow noses that a notched tip that looked triangular in shape, with a fold or notch on it, was the result.

Still, the passion to quantify beauty is not easily quelled. The focus of interest has now narrowed to a particular mathematical ratio, Phi, or the golden section (also known as the golden proportion or divine proportion). Phi was named after Phidias, the Greek sculptor, and is the name given to the division of a line or figure in which the ratio of the smaller section to the larger one is the same as the ratio of the larger to the whole (and equal to 1:1.618). This is supposedly the most aesthetically pleasing point at which to divide the line. Rectangles with sides corresponding to this ratio are called golden rectangles and are believed to be the most pleasing of the quadrilaterals.

Many biological forms show this ratio. We need look no further than our hand to see it: three joints of each of the human fingers bear a golden relation to one another (each is roughly 1.6 times the length of the next, going from the most proximal to most distal from the palm). Once enthusiasts have calipers and protractors in hand they seem to find the ratio everywhere, from shells and flower petals to architectural forms and faces. The pentagram, the symbol of the Pythagorean brotherhood, is composed of line segments all of which are in golden ratio to the segments of the next smaller length. The golden section is found in beautiful music and poetry, including the seven pairs of interludes in the Fugue in D minor by Bach and in the refrain of Voznesensky's poem "Goya."

Gustav Fechner made the golden rectangle one of the first objects studied in the new field of scientific psychology. In 1876 he arranged ten white rectangles of different proportions on a black table and asked people which one was the most aesthetically pleasing. Thirty-five percent of them expressed a preference for rectangles in the golden ratio, and another forty percent picked rectangles close to this ratio. None picked the golden ratio as the least favorite. Fechner then collected data on the dimensions of twenty thousand paintings in twenty-two museums and art galleries to see if great works of art tended to be framed in golden proportions. But the golden ratio did not characterize the relation of height to width of great paintings. For the one hundred and thirty years since Fechner's experiments, no one has been able to agree about whether golden rectangles are beautiful and, if so, why. Many dismiss the claims for the aesthetics of Phi as "numerological fantasies," while others insist they are fragile but real phenomena.

Given the difficulty of testing rectangles, one can imagine how difficult it would be to determine whether golden ratios characterize the complex shape of the human form. Art historian Kenneth Clark lamented that the proportions of Vitruvian man do "not provide any guarantee of a pleasant looking body." As he said, "From the point of view of strict geometry, a gorilla might prove to be more satisfactory than a man" when it came to having limbs that fit neatly into a circle and square.

Several attempts have been made to see whether golden sections characterize a beautiful human face. The most extensive sets of measurements were done by orthodontist Robert Ricketts, who examined the faces of ten attractive models. He discovered golden sections in vertical, horizontal, and depth measurements, and in measures of the underlying skeleton seen in x-ray. He even worked them into the shape and size of teeth. Looked at this way, the face has a peculiarly abstract beauty, as even and harmonious as a honeycomb. Cosmetic surgeon W. Earle Matory, Jr., also finds examples of golden sections in the faces of attractive Asian, African-American, and Middle Eastern men and women.

So far, studies of the golden section have measured only attractive models. None have compared the measurements of attractive faces with those of unattractive faces. It is possible that ratios of 1:1.6 characterize the relations of certain features of the normal human face but do not help us to distinguish the beautiful from the average, or even the beautiful from the plain. There are many points to measure on a human face from the hairline to the brows, eyes, cheekbones, nostrils, lips, chin, and so on. Some distances conform to the golden section, but twist the calipers another way and they do not. We can generate hundreds of indices from the face, and it is natural that some might be in this ratio versus another, especially since Phi is often measured as an approximation. No one has reported a system for the face as a whole.

Phi ratios may be useful for surgeons as best guesses of some pleasing facial proportions (although this will not be clear until they contrast attractive with unattractive faces). But there is no math formula so far that captures the beauty of the human face as a whole. For scientists in this century, the key to understanding human beauty is in our biology, not in mathematics.

Koinophilia: Loving the Average

Studies in the biology of beauty began with the investigations of Sir Francis Galton, Charles Darwin's cousin. Galton was the inventor of fingerprinting, statistical correlations, and an ultrasonic dog whistle (the Galton whistle), as well as an explorer and a eugenicist. In the late 1870s, when Gustav Fechner was spreading his rectangles on tables in Germany, Francis Galton was making composite photographs of criminals in England. His activities had little in common with Fechner's quest for the most aesthetically pleasing shape, but they turned out to have more impact on the study of human beauty. This surprised no one more than Galton.

Galton took images of men convicted of murder, manslaughter, and violent robbery, aligned them at the pupil, superimposed them, and made a single if somewhat fuzzy composite photograph of the faces. For Galton, the individual faces represented variations on a single theme, exemplars of a single visual type. Believing that our minds form general impressions, "founded upon blended memories," he thought his composites were replicas of our mental images.

He was in for a surprise: the composites turned out to be better-looking than the individual faces. In fact, the composites were all attractive. Scrutinizing his "typical criminal" and the faces of the individual criminals, he saw that "the special villainous irregularities in the latter have disappeared . . . the average portrait of many persons is free from the irregularities that variously blemish the looks of each of them."

Galton did not follow up on his discovery of the beauty of blends, probably because attractive criminals did not illustrate any point he wished to make. He traveled by rail with a map in his coat pocket and some pins, composing a "beauty map" of the U.K. Pricking his map whenever he saw a good-looking person, he made London a pincushion but left Aberdeen, Scotland, unpricked. The map did not lead to further scientific discoveries.

But the composite photography he introduced lived on. In the late nineteenth and early twentieth centuries it was popular with college graduating classes, families, and groups of friends who took delight in seeing themselves blended into a single person. As one woman said on viewing a composite, "It is charming to enjoy the society of somebody who is all one's intimate friends at once." We search them to parse resemblance, dis-embed features, and entertain alternate identities. If Galton was right, we are also fascinated by composites because they are scenes stolen from mental life, glimpses of our otherwise invisible mental images.

Scientists can now merge hundreds of digitized images on the computer, and in laboratories in Europe, the United States, and Japan they are using digitized composites to test the beauty of averages. The many people who participate in these studies agree with Galton: averaged faces are usually more attractive than individual faces. Combining two or four faces produces small improvements, combining thirty-two faces makes the composite face much more attractive than the individual faces. Very few individuals are more attractive than the composites (some are, but we will return to this interesting minority later).

Most people do not conjure up the word "average" when they see a good-looking face. But average in this context means average in shape, not average in beauty. In a world of short noses and long noses, almond eyes and round eyes, oval faces and round faces, thick lips and thin lips, overbites and underbites, the eye does its statistics, takes its sums, divides by the numbers, and arrives at a mean value. The beauty of such averages may reflect our sensitivity to nature's optimal designs.

In nature, average proportions often signal good health and good design. Measures of birds killed during storms find a high number with unusually long or short wings. Survivors of the storm have average wingspans that give them the best liftoff and control of flight. Human babies who are born larger or smaller than average (about eight pounds) are less likely to survive. The equation between averageness and healthiness is so strong in the natural world that physiologist Johan Koeslag believes that preferences for the average are stamped into mating animals. He calls it "koinophilia," from the Greek word "koinos," meaning the usual, and "philos" meaning "love."

In 1979, anthropologist Donald Symons proposed the radical idea that human facial beauty *is* averageness. Since the average of a population is likely to reflect the optimal design of physical traits, selection pressures have given us brains wired to calculate means and prefer them. Symons calls this brain mechanism a "face averaging device" and believes it functions like composite portraiture. It collects impressions of faces and turns them into composites that become our standard of attractiveness. We calibrate the looks of every new person we see against this internal composite. Since we evolved in small, isolated groups, each person's composite face was probably very similar to everyone else's. Today each person's storehouse of faces may differ a bit more, making for less agreement about beauty.

Symons made the prediction on the basis of evolutionary biology and the principle that, during most periods, evolutionary pressures operate against the extremes of the population. If this stabilizing selection principle is at work, and people with average physical properties have the best chance of survival, one would maximize fitness by being attracted to partners displaying such properties.

If beauty is indeed averageness, then beauty cannot be a preordained ideal planted in our minds to which faces do or do not conform. The mechanism that stores and averages faces is innate and universal, but the composite it forms is dependent on the faces it sees. This means that in a multicultural world people's internal averages might begin to reflect the universal face, a composite of the features of all races.

Interestingly, the American Academy of Facial Plastic and Reconstructive Surgery has charted the evolution of changes in plastic surgery in America from the 1950s to the 1990s. They note a gradual narrowing of the eyelid from extremely high-lidded in the 1950s (call it hyper Westernized) to increasingly smaller lids in the 1970s and 1990s. The 1950s nose was upturned and narrow with a sculpted tip. By 1990, noses with a wider bridge and fuller tip were preferred. Makeup styles also reflect these changes. The enormous creased eyelids seen on Greta Garbo and other stars of the 1930s are rarely seen today. The favored skin colors are no longer the palest shades. Plump lips, always preferred, are getting plumper. All of these changes reflect an internal average where Asian, African, and Hispanic features are helping to recalibrate norms and reenvision beauty.

Family Resemblance, or Why We Like People Who Look Like Us

Faces close to the average signal healthy and viable potential partners. But certain faces play a bigger role in our lives than in anyone else's and these faces usually belong to family members. This may help to explain why people are often attracted to faces that are strikingly similar to their own. Francis Galton comes back into the picture here. Perusing engagement photographs from local newspapers, he noticed that the betrothed often resembled each other physically, not just in their general level of attractiveness but in their hair color and the shapes of their features. Later studies of engaged and married couples confirmed Galton's observations: they often do look alike. Couples may grow to resemble each other as they assume similar fitness habits and diets, mimic each other's facial expressions, and adopt each other's tastes in fashion. But the resemblance in physical details is there from the start.

Most husbands and wives are similar in many ways: they tend to come from the same religious and ethnic backgrounds, have similar levels of intelligence, and many similar personality traits such as extroversion. But couples also resemble each other in relative height, weight, and hair color, and even in subtle features such as the length of their earlobes and the distance between their eyes. Of course not all couples match up, and most don't match on everything.

The attraction to others who are like oneself is called "assortative mating," and parallels are found in the animal world. The classic studies were done by scientist Patrick Bateson with Japanese quail. Quail chicks were raised for one month with their siblings and then isolated until sexual maturity. Bateson tested their mate preferences by placing them on a stage set consisting of rows of cages and a passageway through the center. Inside the cages were unfamiliar birds (either first cousins or unrelated birds) or familiar birds (the siblings they had been raised with). The quail were allowed to stroll along the passageway and peer in at the cages (Bateson dubbed his stage set the Amsterdam device after the Amsterdam red-light district). Both male and female quail spent longer periods of time gazing at their first cousins than at the other birds.

When Bateson put the birds together in pairs, first-cousin pairs produced eggs three to five days earlier than pairs of strangers or siblings (the other pairs eventually caught up). Bateson interpreted their mate choices as a fine-tuned preference for the moderately new, which he believed might also be true of humans. As he wrote, "Humans may strike a balance between the costs of inbreeding and those of outbreeding and in doing so, may rely on their early experience in a way similar to quail. Clearly many factors deriving from culture and individual experience affect human mating preferences but one influence may well be an attraction towards a member of the opposite sex who differs, but not to a large extent, from well known, closely related members of the family."

Of course people may gravitate to others who resemble family members whether or not they have an optimal outbreeding calculus. The salience of such faces in their internal average will skew their preferences in that direction. Family members may serve as templates for good design. Their faces are indelibly linked with our survival and they are the first objects of our affection. For whatever reason, we do feel a spark for faces like our own. In Hitchcock's *Spellbound,* Ingrid Bergman plays a psychiatrist who explains all of this to Gregory Peck. She complains that poets "keep filling people's heads with delusions, writing about love as if it was a symphony orchestra or a flight of angels." "Which it isn't, eh?" Peck counters. "Of course not, people fall in love as they put it because they respond to certain hair color or vocal tones or mannerisms that remind them of their parents or sometimes . . . for no reason at all."

It has been observed that portraits sometimes bear a resemblance to the artist's face. Perhaps the artist can't help it, the family face is part of his landscape of beauty. Dürer's 1518 portrait of the Emperor Maximilian I is said to resemble Dürer's self-portrait of 1498. It has been suggested that the *Mona Lisa*, which began as a portrait of Isabella, Duchess of Aragon, ended up based on Leonardo's own features. Good portraits must also highlight the sitter's distinctive features to capture the best likeness. Unfortunately, neither the personal intrusions nor the exaggerations of distinctive features may make the sitter look more beautiful to himself. Perhaps this is why painters are often happier than their sitters with the portrait.

Hyperfeminine Females

So far, we have described a beautiful face that looks like the population average, with an overrepresentation of our family members. But if the beautiful is the average, beauty will be, by definition, nondistinctive. To be distinctive a face must have features that are either rare or far from average in shape or size. Such features make a face memorable and eye-catching if not beautiful. Notice how the caricaturist exaggerates what is distinctive in every face, and yet caricatures are often recognized faster and more accurately than an undistorted image and often look more like the person than the person himself. But caricatures usually make the face look less attractive.

If it seems improbable that an attractive face would be one that would blend into a crowd, think about the average model—not a supermodel, but the person who advertises products in the daily papers. It is unlikely that you would recognize the face on the street although you may have seen the advertisements many times. Similarly, beauty pageant winners often appear generic-looking—extremely attractive but not distinctive. The face seems familiar, a better-looking, less irregular version of other faces we have seen. This is part of their appeal. An innate preference for the average may be a way that evolution ensures that human faces rather than other similar-appearing things grab attention. Average faces are the most facelike of faces, and it is perhaps for this reason that it is adaptive for infants to be particularly attracted to them.

But here is the catch: a beautiful face does pop out of the crowd. Some distinctive faces are beautiful. The features of supermodels aren't the norm anywhere but on the runway or the movie screen. Naomi Campbell's and Christy Turlington's lips do not fit population means, nor do Kate Moss's ledged cheekbones and delicate jaw line. As we might suspect, average faces do not score as off-the-scale beauties. On a five-point rating scale, average faces rank between a three and a four, not a celestial five. Average faces are attractive, but they are usually not the most beautiful.

But averageness need not be the only criterion for beauty that natural selection might have favored. When there is competition for partners—the precondition for Darwin's sexual selection—those animals with certain kinds of extreme traits can often be preferred. Such extreme traits, the peacock's tail being the most famous example, can be a sign of the owner's innate resistance to disease and parasites, or an advertisement of its ability to gain sufficient resources to be able to "afford" the flamboyant trait. Any disadvantage of the extremeness of the trait might be offset by the advantage of its attractiveness to potential mates.

Do beautiful faces have extreme traits? Psychologist David Perrett took a large set of images of faces and had them rated for attractiveness. Then he combined the images of the best-looking men and the best-looking women and compared them to the average from the entire group. He found that the composites of the best-looking men and women were more attractive than the composite image of the group. Further, if he exaggerated the ways that the attractive faces differed from the average by caricaturing them, they got even better-looking. Well, this worked for women's faces but not for men's.

But exaggerating any feature in any direction will not make the face more beautiful. In Perrett's study, the most attractive women differed from other women in only a few ways. They had thinner jaws, larger eyes relative to the size of their faces, and shorter distances between their mouths and chins. All of these features exaggerate the ways that adult female faces differ from adult male faces. They also exaggerate the youthfulness of the face. Several other studies using very different methods have come up with similar results.

Psychologist Victor Johnston has a computer program called a genetic algorithm that operates at his website—http//www.psych.nmsu.edu/~vic/faceprints//. Here, users can breed beautiful web offspring. They start by rating thirty randomly selected face images. Then the program "breeds" the top-rated face with another face, creating new faces that replace the lowest-rated faces. By rating rounds of faces, the viewer creates an increasingly beautiful population. According to Johnston, the thirty random faces presented at the start represent seventeen billion points in "face space" (the hypothetical points along which faces may differ). When thousands of people juggle these seventeen billion points, they end up composing a female face with fuller lips, a less robust jaw, a smaller nose and smaller chin than the population average. In one study, the image's estimated age was twenty-four years old, which was two years younger than the population she was derived from. But her lower face had even younger proportions. She had the lips of a fourteen-year-old and the eye-to-chin distance typical of an eleven-year-old.

Cover girls from *Vogue* and *Cosmopolitan* also have larger eyes, smaller noses, and plumper lips than the average "attractive" young woman, and when their facial proportions are fed to a computer, it guestimates their age to be between six and seven years of age. This does not mean that the models actually look like seven-year-old heads attached to adult bodies. But it does mean that the geometry of their facial features is so youthful that the computer, extrapolating its best guess, vastly underestimates their age. Douglas Jones, the author of this study, calls them "supernormal stimuli," women whose attractive features are exaggerated beyond proportions normally found in nature (at least in adults).

Biologist Richard Dawkins created his version of a supernormal stimulus—for stickleback fish—that he calls "sex bombs." The female sticklebacks get swollen bellies when they are ripe with eggs. A crude elongated silvery dummy with a round "belly shape" elicits mating behavior in male stickleback. When Dawkins made the dummy rounder and more pear-shaped, greater lust was inspired. Exaggeration of the pertinent signal worked better than the more realistic depiction. Plastic surgery that enlarges women's lips, raises their brows, and shrinks their noses, in effect, creates human sex bombs.

Babies have almost sexless faces. It's hard to tell boys and girls apart, and many a well-meaning stranger mistakes one for the other. By puberty, the sex differences are fully apparent. During adolescence, testosterone builds up a boy's jaw and chin and brow ridge. Male faces tend to be bigger than female faces, particularly in the lower face. Their faces become more angular and less gently curved as they develop protruding brow ridges and wider jaws. The overhanging brows make their eyes appear more deep set and smaller than female eyes. Because male noses tend to protrude more than female noses, their eyes may appear closer set than female eyes (a small nose, especially one with a depressed nasal bridge, gives the illusion of more wide-spaced eyes). Males have wider noses and mouths than females. It has been suggested that the wider male nose and mouth may have evolved for the more efficient transfer of air to the lungs, to provide the supply of oxygen needed to support the male's higher metabolic rate and his higher hemoglobin count.

Females retain the smoother foreheads and smaller noses of childhood. Their eyes seem bigger and wider because the female forehead is smooth, the eyelashes are longer and stronger, and the eyebrows are thinner and at a further distance from their eyes. Female cheekbones seem more prominent because the female face is flatter, the female nose smaller, and the jaw is less robust so that the face tapers. In a young woman the vermilion border of the lips and the area around it are plumped with fat. As body fat redistributes itself in early puberty, young girls' lips reach their fullest at age fourteen. Compared to a male's lips, their upper lips are delicately curved outward in profile.

Faces with large eyes, high cheekbones, plump lips, small lower faces, and gracile jaws exaggerate each signal of femininity. Victor Johnston believes that the characteristic lower faces of beautiful women, with their plump lips and small jaws and chins, are signs of a female with low androgen and high estrogen. These are women who have gotten through puberty with minimal amounts of male hormones (hence small lower face) and maximal female hormones such as estrogen (plump lips).

When women apply makeup to enhance their beauty, they are heightening these signals as well. They tweeze and lighten their brows to make them seem thinner and farther from their eyes (tweezing is always done from the bottom because the goal is to create not just thin brows but brows farther from the eye). In Greta Garbo's time the brow was completely redrawn. As makeup artist Kevyn Aucoin points out, "Tweezing the brow correctly can make the eyes seem much larger and open up the whole face." The long, strong female lashes are emphasized with mascara, and the whole eye region may be emphasized by outlining and coloring of the lid.

Women emphasize their cheekbones by applying blush. They are instructed to place it on the "apple" of their cheeks (where the cheeks puff up when they smile), which is not where a natural blush would normally form, but where it is perfectly placed to accentuate the cheekbones. Lips are emphasized and sometimes altered in shape with (usually red) lipsticks. There is an arsenal of cosmetic procedures now for increasing the size of the lips, including lip lifts, injections of fat or collagen, and the insertion of Gore-Tex strips. In the early twentieth century, before collagen, women repeated sequences of words beginning with "p" to round and pucker their mouths. Elizabeth Cady Stanton once said that she would not give feminist literature to women with "prunes and prisms expression." In her 1963 *Beach Book*, the young Gloria Steinem admitted to sucking against the heel of her hand: "this makes thin lips full, full lips firm, and fat cheeks lean."

A small lower face is made smaller by big collars and high hair. High hair is much more popular with women (if men grow hair long, they grow it down) because it tends to alter facial proportions in a feminine direction—moving the center of gravity to the top. High collars also shorten the lower face. Women also exaggerate their facial proportions by the way they position their heads. Princess Diana's famous shy smile, with her chin down and eyes up, emphasized and enlarged her eyes, while making her chin and jaw appear as diminutive as possible. I am told by a colleague that it is virtually impossible to photograph a woman in Japan who does not automatically adopt this pose.

Female faces look more like children's faces because they retain the smooth contours of the child, and basic geometry of the child's face. When psychologist Masami Yamaguchi asked people to create a generic image of a child, adult, man, and woman using her specially designed software, people ended up creating very similar images for the female and the child, and for the male and the adult.

Some have suggested that people, men in particular, are responding to "neotenous" features in women, that is, infantile features. Infant features may automatically evoke nurturance, and men supposedly like the looks of helpless and dependent creatures. However, as much as feminine and youthful faces are preferred in adult women, truly babyish faces are not. Psychologist Leslie Zebrowitz has studied "baby-faced" children and adults who have exaggeratedly high foreheads, high thin eyebrows, large eyes, and small noses. People tend to think of them as naive and weak—in a word, babyish. Although baby-faced women are usually considered more attractive than baby-faced men, babyishness and attractiveness are significantly correlated only in infancy. Attractive women's faces may be more childlike but they are not nearly as extreme as baby faces, and they tend to have attractive mature features as well, such as high cheekbones. When scientists Klaus Atzwanger and Karl Grammar asked men to rate pictures of women, the ones rated more attractive were not rated more babylike.

As faces age, all are pulled in a masculine direction. The lower face lengthens and the eyebrows descend to nearer the eye. Cartilage is deposited on ears and noses, making them longer. The upper lip loses some of its subcutaneous fat and flattens in profile. The curvy lips of youth straighten into a widened line. Since men's brows are nearer to their eyes, their noses are larger, their lips are thinner, and their lower faces are longer than females', the aging face looks increasingly masculine. A woman of any age who has small eyes, a relatively large nose, and wide thin lips will look older and more masculine and will be seen as less attractive. To look feminine is to look young. Some scientists believe that our beauty detectors are really detectors for the combination of youth and femininity.

The Mysterious Male

It's a lot easier to address the issues surrounding the female than the male face. This is because we have a clearer idea of what is going on with female beauty. A handsome male turns out to be a bit harder to describe, although people reach consensus almost as easily when they see him.

Sociologist Allan Mazur has been studying not what makes a man attractive per se, but what makes him look dominant. He describes a dominant face as that of someone who seems to be in charge, in contrast to a submissive face, which is that of someone easily controlled. Mazur asked people to decide whether the faces of the men of the West Point class of 1950 were submissive or dominant in appearance, without giving them any instructions about what such a face might look like. People found the task easy, and agreed with one another in their ratings. The dominant faces were oval or rectangular in shape and had heavy brow ridges, deep-set eyes, and prominent chins, features Mazur called "mature." They were usually also described as handsome. The submissives tended to have round or narrow faces, and were more likely to have ears that stuck out. Mazur found that the dominant-looking cadets achieved higher military rank in their junior and senior years and in their profession years later.

Dominant-looking men are successful not only in battle but in bed. In a study of high-school-age boys, the more dominant-looking a boy was, the more sexually active. In another study, behavioral expressions of dominance (such as posture and body position and active as opposed to passive behavior) increased women's ratings of a man's sexual attractiveness although they did not increase their ratings of his likability.

Masculinity is in the facial bone structure, but it is also in the muscles. One feature that seems to add to male attractiveness is a set of powerful-looking masticatory muscles. Think of the muscular jaws of Brad Pitt and Robert Redford. The masseter is a short powerful muscle in the cheek that we use whenever we clench, gnash, or grind. The muscle gives the jaw a squared appearance. Overuse of the masseter can cause it to hypertrophy, a clinical disorder usually found in late adolescent boys who spend a lot of time clenching their jaws and chewing gum. Two dentists call this "the acquired masseteric look," and remarked that it made their patients' faces look attractive (when not chewing gum, of course).

Beards, mustaches, sideburns, and goatees emphasize the maturity and masculinity of a face, since facial hair only appears after puberty, and appears in abundance only in men. Beards also make the face appear more masculine by accentuating the size of the lower face. Psychologist Michael Cunningham suggests that baby-faced men may make themselves more attractive and more powerful-looking by wearing beards. Balding men are often victims of the baby-face effect because their receding hairlines give them the babylike appearance of a large forehead with features set low in the face.

Facial hair has been less abundant in this century than in centuries past (except in the 1960s), partly because medical opinion turned against them. As people became increasingly aware of the role of germs in spreading diseases, beards came to be viewed as repositories of germs. Previously, they had been advised by doctors as a means to protect the throat and filter air to the lungs. But by 1907 a Parisian scientist was walking the city with two men, one with and one without a mustache, to test the health hazards of the former. After their walk through the Louvre and other sites, they each kissed a woman whose lips had been sterilized. The residue was wiped off and dipped in a sterile solution and left standing for four days. The residue from the clean-shaven man contained merely harmless yeast, but the residue from the mustached man was "swarming with malignant microbes . . . diphtheria, putrefactive germs, minute bits of food, a hair from a spider's legs and other odds and ends."

Beards never quite recovered. In a 1904 article in *Harper's Weekly*, a columnist lamented the aesthetic effects of seeing every male face clean shaven: "The revelations are sometimes frightful: retreating chins, blubber lips, silly mouths, brutal jaws, fat and flabby necks, which had lurked unsuspected in their hairy coverts now appear. . . . Good heavens, he asks himself, is that the way Jones always looked?" In 1982 columnist Otto Fredrick shaved off his mustache. "To my dismay I saw in the mirror a face that I had not seen for more than a decade and I hardly recognized it. How had I acquired those deep vertical lines of discontent across both ends of my mouth?" Beards and mustaches have been used by men in much the same way that women use makeup to camouflage unattractive features and to mask the signs of aging. A well-placed mustache can make an asymmetrical nose or mouth look more symmetrical, and beards and goatees can change the apparent shape of the chin and jaw line. In psychologists' studies, beards have a significant impact on face recognition.

It would appear that emphasizing all the characteristic male features would make a man more handsome. If we look at magazine ads, we seem to see chiseled faces, all brows and jaws and piercing, narrowed eyes looking back at us from advertisements. In Botticelli's *Young Man*, painted in the 1480s, he exaggerated the male characteristics of the young man's handsome face by painting the different areas of his face as if seen from different angles. The artist appears to look straight into the boy's eyes and lower lip, but views his chin, nose, and brow ridge from below.

But some psychologists have said there may be limits to just how manly we want a man's face to look. Psychologist Michael Cunningham suggests that women have "multiple motives" when searching for a mate. She will want evidence of his prowess and resources, but she will also want evidence that he will invest those resources in her and a potential child. A face that conveys dominance may not provide her with cues to dependability or desire to invest. In Cunningham's studies, women are attracted to typically masculine faces with some nontypical features. For example, women like men with large eyes and men with wide smiles (smiling per se does not increase attractiveness, but certain kinds of smiles are very attractive). Cunningham suggests that large eyes are a "neotenous" cue, one we associate with babies, and one that elicits female nurturance. In general, however, he found that the more masculine a man's face was perceived to be, the more women rated him as a desirable and physically attractive date or mate. Men with baby faces were rated as less desirable for marriage, less masculine, and less attractive.

We have seen that exaggerating the femininity of female faces makes them more attractive. However, exaggerating the masculinity of male faces may make them less attractive. Psychologists Tatsu Hirukawa and Masami Yamaguchi created "hyper" masculine and "hyper" feminine faces that exaggerated the ways that male and female faces differ from each other. They found that women rated other women's faces as most attractive when they were average in shape, but that men preferred the hyperfeminine face. Both sexes found the hypermasculine males significantly *less* attractive. David Perrett and his colleagues found that men and women in Japan and Scotland judged "hyper" feminized female faces more attractive, and "hyper" masculinized male faces significantly less attractive. Indeed, they found that the most attractive male face shapes were those that had been slightly feminized or "softened."

Hypermasculinity may suggest certain undesirable personality traits. For example, in one study, men with high levels of circulating testosterone had faces that were judged higher in strength and dominance but lower in goodness and friendliness than other men's faces, regardless of their expression. These men also had smaller smiles. In Perrett's study the hypermasculinized faces were also perceived as more dominant and masculine but less warm, honest, and cooperative, and lower in potential quality as parents.

Surgeon Paolo Morselli reports an interesting case from Italy of a thirty-eight-year-old man with no psychiatric history whose face looked so fierce and aggressive that it was causing him serious social problems. Morselli dubbed this problem the "Minotaur syndrome" after the Greek myth of the half monster, half man who was Dante's symbol of brutality. The man got plastic surgery to change "a threatening face into an inoffensive face." His jaw and brow ridge were both reduced, and the man said that people were friendlier and kinder to him afterward, and he felt more himself.

The Perrett study is the only empirical evidence to date that some degree of feminization may be attractive in a man's face. All the others suggest that, for men, the trick is to look masculine but not exaggeratedly masculine, which results in a "Neanderthal" look suggesting coldness or cruelty. Michael Southgate helps create mannequins, and he says that male mannequins are always much more difficult to get right. "You can sell five hundred female mannequins to a store, with no problem. But you deliver ten males, and everyone from the managing director down to the elevator operator has something to say about them. "This one looks like a rapist This one is a murderer. This one is a fag." If a face looks too feminine, the manufacturers assume it is a face that will be loved by men not women. If it is too masculine, it looks criminal.

When a large group of men and women were asked what characteristics they would like to show in a photograph, many spontaneously came up with the combination "beautiful-handsome-smart-friendly-kind-nice." This combination may be easier to achieve for females than it is for males.

Fleeting Seduction

The face also has its fleeting signals—smiles, frowns, invitations, and refusals. "I'm in heaven when you smile," Van Morrison sang, because happiness is infectious. But it doesn't necessarily make you better-looking. A woman's smiles may enhance her beauty slightly, particularly if she has white, even teeth. It appears that high-testosterone males are better off not smiling, since their smiles may make them appear less likable (perhaps the origin of poker-faced action heroes like Clint Eastwood). The big wide smiles such as those shown by actors Tom Cruise and Matt Damon are very attractive.

What does seem to affect beauty is a look of potential receptivity. When we are excited our pupils dilate automatically, independently of light conditions. When photographs are retouched to increase the size of a woman's pupils, men think that she looks more attractive, although they are not aware that this is the basis of their response (they think she looks "more feminine" or "prettier" but none noticed that she had larger pupils). The same thing happens when women look at pictures of men. They too find the man with enlarged pupils more attractive. Men are more likely to volunteer to be a woman's partner in a psychology experiment, and women are more likely to volunteer to be a man's partner when the potential partner's eyes are dilated pharmacologically. Like the aperture in a camera lens, opening up the eye lets in more light but allows less depth of field. The dilated eyes are giving us the kind, gauzy look of the soft-focus camera.

When we are sexually excited our lips get redder, swollen, and more protuberant, just like our nipples. Desmond Morris believes that the shape of our lips has evolved to highlight this signal, and that female cosmetics exaggerate it. Other primates have lips that protrude and can roll back temporarily, but only the human has lips that constantly expose their pinkish-red mucosa. Noting that many animals have upper-body features that mimic their genitalia (for example, male mandrills have red faces and blue cheeks that mimic their red penises and blue scrotums), Desmond Morris suggests that female lips are a "labial mimic."

Although the face may not be so literal a parody of the sexual organs, its signals of potential sexual interest up its attractiveness. Even when fully dressed women are photographed in fashion magazines to be viewed by other women, their subtle gestures of sexual provocation make their beauty more appealing. As Alexander Liberman, formerly of *Vogue*, said, "The good model is . . . involved in provoking the photographer—by her movement, her expression, her attitude—to fall in love momentarily and to capture this fleeting seduction."

Symmetry

When photographer Andre Kertesz photographed Mondrian in his studio he noticed that the great painter of grids had trimmed his mustache to make his face look symmetrical. Like Mondrian's, most faces have some minor fluctuating asymmetries. This is why people like to be photographed from one side or another, usually to mask the side with the "flaw." Marilyn Monroe was virtually always photographed from her right side. It is said that King Edward VIII (the late Duke of Windsor) had such a strong preference for revealing the left side of his face that he objected to the coin designs produced for him because they showed his right profile.

Fluctuating asymmetries are random deviations from perfect symmetry, for which the mean asymmetry is zero. For example, we assume that eyes and wrists and breasts are the same size and, if not, that they will vary randomly across people. Features can deviate from perfect symmetry for many reasons, including exposure to pollutants and parasites during development, and malnutrition or disease. As people age, facial asymmetry can also increase. Biologists Randy Thornhill and Anders Moller have found that in many species asymmetrical animals have lower survival and growth rates, and diminished reproduction.

If symmetry is an indicator of health and fitness, like averageness, it should also appear attractive to us. But since average faces tend to be symmetrical, it's been difficult to tease the two effects apart. It does appear, however, that facial symmetry per se is attractive in both men and women. But not all studies have found perfectly symmetrical faces attractive, and they help us to appreciate many of the natural and not unattractive directional asymmetries of the face.

Directional asymmetries are not random and fluctuating but marked across the species—handedness is an example of a directional asymmetry. Our facial expressions, controlled largely by the right side of our brains, tend to be more exaggerated on the left side of the face. This is particularly true for deliberate or controlled expressions. When we talk, we tend to move the right side of our mouths more than the left, probably because it is the left side of the brain that controls speech.

When scientists create symmetrical faces by making mirror image composites of left-left or right-right faces, these faces may bear unnatural expressions. For example, one composite may look virtually expressionless, while the other may have an unnaturally intense and symmetrical expression. If midline facial structures (such as the nose) deviate slightly, each composite will have features that do not correctly convey their shape. For these reasons, composites made from blended whole-face mirror images (each side is a composite of both the left and right) without expressions are the only ones that can test whether symmetrical faces are more attractive. When this method is used, symmetrical faces are found to be more attractive. Of course, a face can be symmetrical and not beautiful. On a one-to-ten scale, symmetrical faces tend to rate from six to eight, not ten. Like averageness, symmetry is an ingredient of the attractive face, but no guarantee of stunning beauty.

The B Spot

Babies and adults automatically recognize beautiful faces. People make snap judgments about appearance all the time. They tend to agree with each other about who is beautiful, and they tend to be guided in their judgments by mechanisms that detect symmetry and averageness as well as exaggerated markers of femininity in women's faces. This suggests that the general geometric features of a face that give rise to the perception of beauty may be universal, and the perception of these features may be governed by circuits shaped by natural selection in the human brain.

Victor Johnston, a psychologist at New Mexico State University, did a pioneering study in which he put electrodes on people's scalps to see what happens to the brain's electrophysiology when we look at faces. Johnston finds a relation between what is known as the LPC (the late positive component of event-related potentials) and the sight of the most attractive female faces. LPCs normally fire in response to stimuli that have "affective value," that is, command attention and pack an emotional wallop. It is these electrophysiological signals that show greater amplitude when people look at beautiful female faces as opposed to plain faces. Johnston's study shows us that beauty's frisson is demonstrated in the brain—arousal and riveted attention are detected in our electrophysiology as we gaze at a beautiful woman.

We also know that faces are recognized faster and more accurately in particular regions of the brain's right hemisphere (in the temporal and occipital lobes). In my research with patients who had suffered strokes to regions of the right or left hemisphere, I found that only the right-hemisphere-damaged patients showed marked difficulty in recognizing human faces and facial expressions of emotion. The brain's asymmetry in recognizing faces has some interesting consequences. For example, the right side of the face seems to "resemble" the whole face more than the left side. If we compose mirror image composites, it is the one composed from the right side of the face that seems to better capture the likeness. This led to a flurry of speculation, including that we might reveal a public self on our right and a hidden self on our left. Some psychologists proposed that the degree of facial asymmetry could reveal the extent of neurotic conflict within the individual between the public and private selves!

But the explanation turns out to be rooted in our perceptual wiring. Since visual routes cross in the brain, whatever is seen in the left hemifield gets to the right hemisphere first. When we face another person, the right cheek projects directly to our right hemisphere, while the left cheek arrives there later by a slightly more circuitous route. The features on the right side of the face are processed first by the brain and they make that side of the face predominate in our minds. This is why two right sides look so much like the whole person. It is also why people like faces better when they are viewed in the familiar way. Individuals prefer to see their own face reversed, because that is the way they are used to seeing it (in a mirror), while their friends prefer photographic images of them.

Whatever projects to our right hemisphere also dominates our judgments of the face's attractiveness. Psychologists David Perrett, D. Michael Burt, and their colleagues created blended chimeras of faces whose halves differed in levels of attractiveness. They found that when the attractive half of the face projected to the right hemisphere people saw the whole face as attractive.

How the brain recognizes and responds to beauty is one area of research in my laboratory at Massachusetts General Hospital and Harvard Medical School. My colleagues and I have been tracing the neural pathways by which we recognize one another's faces and facial expressions of emotion. We have focused our attention on brain regions involved in high-level vision and emotional learning, particularly the right temporal lobe and an underlying subcortical structure called the amygdala. We have found evidence for discrete and localizable brain circuitry involved in the recognition of faces, and in the recognition of certain facial expressions such as fear.

In related work, my colleagues Hans Breiter, Steve Hyman, and Bruce Rosen have been mapping the brain circuitry involved when we experience pleasure and reward or when we crave a previously rewarded state. We are using these efforts as the framework within which to make predictions about the circuitry involved in our automatic and pleasurable response to human beauty. If detection for beauty is hard-wired, governed by circuits created by natural selection, we should be able to uncover its distinctive pathways. Using brain imaging technology, we are now watching the brains of men and women as they stare at faces. The subjects for our studies are heterosexual and homosexual men and women, and the faces they see are either strikingly attractive men or women, or men or women average or below average in attractiveness. This research is in progress and it is too early to report our results, but we do have some intriguing evidence to report from related research.

Although the presence of right-hemisphere advantages for so many face skills suggests that our right-hemisphere face analyzers may be doing double duty when recognizing a face's attractiveness, this may not be true. For the past ten years I have been conducting an intensive case study of a man in his mid-forties who suffers from the rare syndrome called prosopagnosia. Prosopagnosics are unable to recognize people by face, including their children or their own faces in a mirror. This man's wife has to wear a special ornament—a ribbon of a certain color or a distinctive hairclip—when they attend public events so that he will be able to find her. As I drove him home once, I saw two children in his driveway. I asked him if they were his children and he replied, "Must be, they are in my driveway."

This man can recognize people by voice, by their perfume, and even by their gait. His inability to recognize faces is a result of a head injury he suffered in his first year at college, which extensively damaged parts of his right hemisphere. Although he was able to graduate from Harvard, attain two master's degrees, marry, have children, and work as a full-time professional, he has never been able to regain his ability to recognize a single human face. This holds true even though he can recognize almost everything else (although he also has trouble distinguishing among some four-legged animals and among some facial expressions). It's only with individual faces that he draws a complete blank.

I was intrigued to hear this man describe certain people as attractive, and wondered if he truly saw faces the same way that others do on this dimension. In a battery of tests, he rated people's attractiveness very much the way everyone else does. He shows sensitivity to symmetry of faces, to slight caricature effects (exaggerating the differences between beautiful and plainer faces), and to increasing the femininity of female faces, although all of this requires attention to subtle details in the face as well as to the configurational or gestalt properties of the face. He may not know who a person is, but he knows when he finds her attractive. Natural selection has kept these two circuits at least partially separate. Beauty, after all, may travel on its own pathway.

CHAPTER SIX

Size Matters

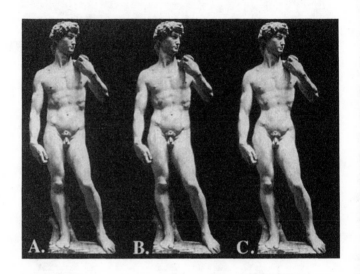

Man admires and often tries to exaggerate whatever characteristics
nature may have given him.

—CHARLES DARWIN

I worked with Freud in Vienna . . . we broke over the concept of
penis envy. Freud felt it should be limited to women.

—WOODY ALLEN

Too much of a good thing can be wonderful.

—MAE WEST

The body of a male animal is sculpted by his need for the female. To fight off competitors he develops canines or antlers or grows massive in size. To charm the female, he displays beauty. Birds are lovers, not fighters, and they are "ornamented by all sorts of combs, wattles, protuberances, horns, air-distended sacs, topknots, naked shafts, plumes and lengthened feathers gracefully springing from all parts of the body. The beak and naked skin about the head, and the feathers are often gorgeously colored." The description is by Darwin, who was dazzled by our fine-feathered friends and guessed that female birds were too. After all, he wrote, "strong affections, acute perception, and a taste for the beautiful" are not incompatible with "low powers of reason."

Beauty is not essential for reproduction, but it helps. The "unarmed, unornamented, or unattractive males would succeed equally well in the battle for life and in leaving numerous progeny," Darwin said, "if better endowed males were not present." But they are, and they whisk away the mates.

Females prefer flamboyance to subtlety. Female swordtail fish like males with long swords (colored extensions of the caudal fins), female swallows and widowbirds prefer males with long tails, and so on. But flashy ornaments have a cost to the male. The peacock who lugs his train of one hundred and fifty to two hundred iridescent eye-spattered feathers has, as Helena Cronin writes, "not only a wife and children to support, but wife, children and a tail!" The ornament attracts the attention of predators, uses up precious metabolic resources, and has to be carted about.

But the females reward the males for their trouble. Peacocks with the most elaborate trains (most eye spots) not only have more mating success than other males, they snare virtually every peahen. Great snipes whose white tail patches are brightened with Wite-Out and male swallows whose tails are fitted with glued-on tail extensions have more mating success than normally endowed males. This suggests that females prefer traits that are exaggerated beyond what is found in nature.

The knotty question is why females favor extravagance. In 1930 geneticist Sir Ronald Fisher described an explosive, escalating "runaway" process of sexual selection that could account for it. "Runaway" works like this. A preference is expressed, for example, when a female bird takes a fancy to a male bird with a big tail. The female mates with her relatively big-tailed male and they beget birds with big tails. The preference spreads as new females take notice of the big-tailed birds. Once a preference develops for whatever reason, there is a selection pressure to conform to that preference because only then will the females have sons who are preferred as mates. This pattern repeats itself and, over time, females no longer want to mate with smaller-tailed males. The preferences grow stronger, and the tails get larger.

Large and showy displays may be the products of such evolutionary runaways. What were once adaptive traits that eventually spiraled into a runaway may now be sustained by popularity alone, much the way a best seller, a hit record, or a fashion trend takes off. But most biologists think that irresistible beauty in biological forms is about more than popularity. They argue that beauty is not arbitrary or capricious but a form of communication.

According to one theory, put forward by biologist Amotz Zahavi, showy traits are a form of "handicapping." The peacock lugging around its huge train is saying, I am so healthy and strong that I can afford my tail of sixty inches in radius, can siphon off nutrients to keep it in brilliant shape, and allow myself to be vulnerable to sneak attacks from behind. Only an animal in prime condition can develop and carry such displays, raising immune defenses against parasites to keep it lustrous and abundant. The signal is kept honest and reliable by its costs.

The female's interest in this prime physical specimen is fueled by her desire to have viable offspring. Do males with fancy ornaments make the best fathers? While there is no evidence that they lavish better care on their offspring, they do seem to pass on hardier genes. Peacocks with elaborate trains sire offspring with a better chance of survival than peacocks with less flamboyant tails. Barn swallows with longer tail ornaments have better long-range survival than swallows with shorter tail ornaments, and tend to have long-lived offspring. Red-throated three-spine sticklebacks are less likely to lose eggs from the nest through predation than their drab companions. The red throat functions as a threat to other males and a charm to females—suggesting that some showy traits are effective in both love and war. Flies of the diopsidae family have eyes on long stems, some of which are longer than the entire length of their bodies. Females prefer males with the longest stems, and it turns out that these males have a hidden benefit, a "tough Y" chromosome that increases the chance of producing sons in a population that is overwhelmingly female.

Big ornaments also tend to be more symmetrical than small ornaments and, as we will see, females prefer symmetry. Symmetrical traits are difficult to produce, especially if they are large and subject to sexual selection. But if large ornaments are a handicap, an honest advertisement of genetic quality, symmetry is exactly what you would expect. Animals that can develop large ornaments should be better able to buffer developmental stresses than small animals. Since secondary sexual characteristics are maintained by circulating hormones, which compromise immune function, only the most fit animals can maintain them and retain immunocompetence.

In their natural state, male swallows have tail feathers that are about twenty percent longer than the tail feathers of the females. The males with the longest tails attract more mates than other males. But their long tails are also more symmetrical than the shorter tails. Zoologist Anders Moller wanted to know whether tail length and symmetry independently affect female choice. To find out, he performed minor cosmetic surgery, snipping and gluing tail feathers onto male birds to make the tails shorter, longer, symmetrical, and asymmetrical. To be sure that glued-on feathers per se were not of interest, Moller also snipped and re-pasted the feathers of other males without altering their size or symmetry. Females cared about both size and symmetry: males with long but asymmetrical tails were not as popular as the males with long symmetrical tails.

Zoologists Eystein Markusson and Ivar Folstad found that the biggest antlers among rutting reindeer were also the most symmetrical. Interestingly, the relation of size and symmetry did not hold for other body parts. Markusson and Folstad suggest that the conditions of the antlers give specific information about parasite burden, which other parts do not reveal. Sexual selection may have favored the evolution of display traits that easily reveal asymmetry such as antlers, particularly the asymmetry caused by a compromised immune system.

But what do the mating preferences of peacocks, swallows, snipes, and reindeer tell us about human beauty?

Tower of Power

During the O. J. Simpson trial, defense attorney F. Lee Bailey asked if he could enter as evidence a leather glove in a plastic bag, in order to illustrate how a detective could have picked up a glove at the crime scene and hidden it. Prosecutor Marcia Clark objected that the real glove was a size extra large, while the glove Bailey was trying to put into evidence was a different size. "Size small," she said sarcastically, "I guess it's Mr. Bailey's." She could not have uttered a more withering remark. Likewise, when Tolstoy wanted to convey his scorn for Napoleon he described him as a small man with small hands.

In the animal world, the dominant animal tends to be the largest. In human villages throughout the world, the chief is known as "the big man," and he is usually physically imposing. Our first President, the six-foot-two-and-a-half-inch George Washington towered over his peers, and so have his successors (since 1776 only James Madison and Benjamin Harrison have been below average height). The easiest way to predict the winner in a United States election is to bet on the taller man: in this century you would have had an unbroken string of hits until 1968 when Richard Nixon beat George McGovern.

Men who are above average height often acquire more resources than other men. The average height of a man in the United States is five feet nine. More than half of the CEOs in American Fortune 500 companies are six feet or taller and only three percent are five feet seven or less. Height may be an unacknowledged part of the package that companies look for—they want someone who can not only do the job, but who fits an image, who can literally be looked up to. When corporate recruiters were asked to choose between two applicants who were matched on all qualifications except body size (one man was six feet one, the other was five feet five) seventy-two percent of the recruiters chose the taller applicant, and only one recruiter preferred the shorter man (the rest did not have a preference).

Examining the employment patterns of over a thousand graduates of the University of Pittsburgh Master's in Business Administration program, psychologist Irene Frieze found that height had a significant impact on starting and current salaries. Her study was done in the mid-1980s and the average salary of her respondents was forty-three thousand dollars. But the six-footers were earning over four thousand dollars more than their five-foot-five-inch counterparts. In Frieze's study, height did not have a significant impact on the women's salaries. A study of young adult men and women in a wide range of jobs found significant effects of height for both men and women, although the effects were larger for men. The advantage to the tall would not be discriminatory if tall people outperformed the short on the job. But these studies are of office workers and executives, not basketball players, and there were no differences in the quality or quantity of work performed based on height.

Shorter men also face difficulty negotiating the social hierarchy, particularly as children. Writer Stephen Hall was very short as a child, "a time in life when logic and verbal dexterity were not the preferred means of conflict resolution I was . . . punched, pummeled, bumped, slapped, shoved, thumped, dumped, ridiculed. . . . Part of it was garden-variety childhood bonhomie, true, but some of it had a nasty predatory edge. I don't associate the abuse with tall people, necessarily; just people taller than I, all of us bit players in some great trickle-down theory of torment." When men of short stature act tough, they are accused of having a "Napoleon complex," a desire to compensate for physical stature with excessive need for power. As Ian Fleming said in *Goldfinger:* "Bond had always mistrusted short men. They grew up from childhood with an inferiority complex. All their lives they would strive to be bigger than others who had teased them as a child." With age, humor, intelligence, and talent a short man can stand above a tall man, but not before dealing with their unthinking presumptions.

It is no wonder that we are always trying to create the illusion of height, whether we possess it or not. William III had door knockers placed above eye levels at Hampton Court in London to make people who visited feel short (and see him as tall). Speakers mount stages, religious leaders speak from altars, kings and queens ascend thrones. John Wayne was six feet four but, according to Robert Mitchum, "he wore four-inch lifts and a ten-gallon hat. . . . He even had the overheads raised on his boats so he could walk through the doorways with the lifts on." And if all else fails, people convince themselves and others that they are just a bit bigger: seventy-one percent of the time people slightly overestimate their height.

The unconscious association of power, status, and height is so ingrained in us that we automatically presume that powerful people are big people. In one test of this idea, a man comes to lecture to students and is introduced as a person either low or high in the dominance hierarchy of his field—from a fellow student to a senior professor. After the lecture, the students are asked to guess the height of their visiting lecturer. They guess that the professor is several inches taller than the student, even though the same man plays both roles.

All of this would be irrelevant to beauty if it did not make a man more sexually attractive. But it does. The cliché description of the attractive man is "tall, dark, and handsome." In one study of personal ads, eighty percent of the women who mentioned height wanted a man who was at least six feet tall. Tall men get more responses to personal ads whereas height does not influence the popularity of a woman's advertisement. In her study of stereotypes, psychologist Linda Jackson found that taller men were assumed to be more athletic, masculine, physically attractive, and higher in professional status than shorter men.

There is some disagreement about exactly how tall the ideal man should be. Most studies find that women like men in the six-foot to six-foot-two range, the height of most models. But in one study women preferred men on the tall side of average (five feet nine to five feet eleven). All agree, however, that men at or above the population average are preferred to men who are below average in height. Of course, tall or short is always in comparison to local norms. In the coastal town of Accra in west African Ghana, a five-foot-five-inch man is called "Big Joe." Recent immigrants from the interior who are six feet five are teased about their stature.

Men everywhere are, on average, taller than women. One would expect pairings between tall women and shorter men to be uncommon but not rare. Yet, in one study of married couples, less than one percent (.3%) of women were taller than their husbands, a percentage considerably less than would have been expected by chance. Although people presume that the "male taller" rule in mating is a product of males' desires to tower above their companions, it turns out that both men and women care about the man's height. Even women choosing among the profiles of potential sperm donors in a clinic show a preference for a tall donor! Clearly, height is seen as an important trait to pass on to a child.

Height is determined by many factors, including climatic conditions, diet, and genes. The genes set the upper limit of growth, but its acquisition is determined by life conditions. With better nutrition, height has steadily increased in this century, particularly for the upper classes, who have been taller than the lower classes in every country since the late eighteenth century. Relatively greater height within a population suggests tall ancestors who garnered a good share of the resources, and a good food supply for the young when their bodies are developing. Shorter stature is associated with hunger, food intolerance, and illness. In general, taller individuals have better reproductive histories.

Despite all the advantages of height, a tall woman was once considered socially handicapped because she had a smaller pool of potentially taller mating partners. Tall women wore flats, slouched, and did what they could to deemphasize their unusual stature. It may seem hard to believe but during the 1940s over five hundred young women were given high doses of estrogen to attempt to halt their growth. A psychologist's survey in 1959 found that half of the women wanted to be shorter (and all but one man wanted to be taller). Today, many of the most famous female beauties are tall, including the six-foot Brooke Shields and Uma Thurman, the five-foot-ten-inch Nicole Kidman and late Princess Diana of Wales, and the five-foot-nine-inch Cameron Diaz and Gywneth Paltrow. The average model is just over five feet nine. In the United States, a five-foot-ten-inch woman is taller than fifty percent of men, and a six-foot woman is taller than eighty-two percent of men. Does this mean that the "male taller" rule is relaxing? Perhaps, but most tall women end up with tall men. Noticeably shorter husbands are usually very rich and powerful (Aristotle Onassis, Prince Rainier, Henry Kissinger, Tom Cruise).

No Pecs, No Sex

Written on a billboard advertising David Barton's gym in New York, NO PECS, NO SEX could come straight out of studies of tail lengths and peacock plumes. Pectoral muscles are the human male's antlers, their weapons of war. Men may no longer hunt or wage war through hurling weapons but a broad chest still echoes of survival skill. Men are fist fighters who flex muscles instead of baring canines. Today a man's upper body strength is often used for a different form of hunting and gathering. Six million people read *Muscle and Fitness* magazine each month, where the typical cover shows a muscular male (often being gazed at by an admiring female) beside headlines such as "Getting Big."

Studies that focus on weight suggest that heterosexual men are happy with their bodies, or at least less dissatisfied than women or gay men are. If they looked at shape, height, and muscularity, they would find a less happy group. Many men consider themselves underweight. But they don't want to get fatter, they want to get more muscular. Men have less body fat than women (twenty-five to twenty-seven percent body fat is the average for women, versus fifteen percent for men) and have no interest in acquiring more. But they are eager to add muscle mass as early fitness advocate Charles Atlas knew when he promised to transform weaklings into "he-men."

Generally, the most attractive male torso is thought to be V-shaped, tapering from wide shoulders to a narrower waist and hips. The most strongly disliked shape for men, according to both females and males, is the pear shape, with thin shoulders and widened middle and bottom. In one study Caucasian and Japanese male students all expressed desires to be larger everywhere except around their hips and waists.

Adult men have, on average, more muscle mass than women and the biggest difference in strength is concentrated in the arms, chest, and shoulders. The average woman can bench-press about a third of the weight that a man can, and her grip strength is about half that of a man's (contrast this to leg strength, where the average woman's strength is about three quarters that of the average man). The average male gym attendee's first goal is to exaggerate that difference by developing his upper body—the pectoral muscles in the chest, the lats (latissimus dorsi muscles) in the back, and the biceps and triceps in the arms. Arnold Schwarzenegger was said to be able to balance a glass of water on his flexed pecs. When actor Tom Hanks asks his friend Rob Reiner what women are looking for these days, in the 1993 movie *Sleepless in Seattle*, Reiner answers, "Pecs and a cute butt."

In Elizabethan times men stuffed and padded their doublets to get that Calvin Klein ad look, much as Roman warriors wielded huge breastplates and as men today pad the shoulders of their jackets. When leg shape was visible in tights, before the advent of pants, men went about stuffing those too if their legs were thin. When the eighteenth-century cavalier high boots went out of style, thin-legged men wore two pairs of tights with calf pads tucked inside.

Current fashion for both men and women displays the body rather than camouflaging it. Those not genetically gifted or without the time or interest to spend hours in gyms are taking shortcuts such as liposuction and implants. There are pectoral implants to make the chest appear broader and more muscular and calf muscle implants to give legs more heft. There is also liposuction to remove flabby stomachs and breasts, two big detractors from the male chest-of-armor look. Medieval men wore their padding on the outside, some modern men wear it beneath the skin.

Male models have bodies almost as uniform as those of female models. They are tall—six feet or taller—and have V-shaped torsos (the usual measurements are a forty- to forty-two-inch chest and a thirty- to thirty-two-inch waist). Mannequins, too, are six feet two with forty-two-inch chests. The silhouette gets increasingly exaggerated as we move from high fashion to fiction. Fabio, who adorns the covers of romance novels, has a forty-four-inch chest. Cartoon character Batman has shoulders that have morphed from one fourth of his height to almost half of his height. Even Al Capp's Li'l Abner, who started out in the 1930s as Mammy and Pappy Yokum's slim country boy, had become, by the 1950s, a guy with big biceps and shoulders and a narrow waist.

Male body builders present the extreme manly silhouette—cut to eight percent fat but massive with muscle, the shoulders almost twice as wide as the waist (which is not unlike the cartoonish female with the eighteen-inch waist and the thirty-six-inch bust). Arnold Schwarzenegger, five times Mr. Universe, has a fifty-seven-inch chest and a thirty-one-inch waist, and his neck, arms, and calves are all about the same size (eighteen to twenty inches). Sylvestor Stallone has summed up the increased attention to the beauty of the male physique by predicting that "the guy with the eighteen-inch arms, the thirty-one-inch waist, the male-model chiseled, Calvin Klein ad type of person, he is, for the nineties, the woman with the triple E. He's taking the place of the blond bombshell of the fifties." As one journalist wrote, "If fat is a feminist issue, then muscle is a masculine issue."

And as some women have developed an unhealthy obsession with body fat, some men (and a small number of women) have developed an unhealthy obsession with muscle. Psychiatrist Harrison Pope has done a series of studies of athletes and body builders who attend gyms in Boston and Los Angeles. He finds that about nine percent of them suffer from "a reverse form of anorexia nervosa," which he has named muscle dysmorphia. The syndrome is characterized by a person's distorted perception that his body is weak and small when it is in fact large and muscular. Just as the anorexic starves herself because of her fears of getting fat, the patient with muscle dysmorphia may abuse steroids and use food supplements for fear of not being big enough. Half of the people Pope interviewed for his studies used anabolic steroids (drugs that stimulate the production of cellular protein and aid in muscle growth). Pope believes that muscle dysmorphia belongs among the "affective spectrum disorders," which include anxiety, depression, obsessive-compulsive, and eating disorders. It may be more prevalent now because of the ease with which people can obtain steroids, and because there is greater social pressure on men to have muscular bodies. Although muscle dysmorphia may be a rare psychiatric disturbance, steroid use and the desire to bulk up are not. Two recent studies of high school students found that six to seven percent of high school boys are using or have previously used steroids.

Penis, Threat or Charm?

Now, let's get back to Mr. Bailey's evidence—O.J.'s glove. On CNN's "Crossfire" that night, Barry Tarlow accused Marcia Clark of making "jokes about the size of F. Lee Bailey's penis." Another commentator perked up, saying, "What, I missed a colloquy on a penis?" Mr. Tarlow answered in a world-weary tone, "Come on . . . I don't think you need to be too swift to pick that up. When I was young, condoms were referred to as gloves and I think it's an obvious sexual reference to the size of his penis."

Americans are obsessed with large sex organs—be they women's breasts or men's penises. In the 1930s mannequins were imported from Europe and came in size small, medium, and American according to the size of the genitalia. But Americans aren't the only ones who like to exaggerate. Phallocrypts, part of the male attire in New Guinea, are sheaths that cover the penis and they run to two feet in length.

The penis is considered the reflection of a man's prowess. In most surveys men seem to feel that theirs are not large enough. Penis size and shape are one source of concern for male patients with body dysmorphic disorder or imagined ugliness. As one patient said, its size made him feel "half a man . . . my penis doesn't look ugly but it looks unattractive and unmanly."

Penis size has little to do with the size of the rest of the body, and a tall bulky man can have a smaller penis than a short thin man. The archived data from the Kinsey Institute of Sex Research show that the average penis of a college man is three to four inches when flaccid and five to seven inches when erect (with the range for the erect penis being 3.75 to 9.6 inches). A recent sample of over a thousand men between the ages of twenty and sixty-nine years old showed that the Kinsey data were still basically correct although there were a greater proportion of shorter erections with lengths in the 4.5 to 5.75 range representing forty percent of the total.

Man's concern with penis length is ironic—compared to most other primates, he is gigantically endowed. Four-hundred-and-fifty-pound gorillas commanding harems of females have penises that are a little over an inch when erect. Among the great apes only chimps have proportionately bigger testes than humans (but not bigger penises).

But we humans have the capacity to create supernormal stimuli, and we use it. The simplest way is to pad. In the movie *Spinal Tap*, a hilarious send-up of a rock group, the bassist was stopped by an airport metal detector for carrying a foil-wrapped cucumber in his pants. Today, over ten thousand American men have tried a more permanent solution: surgery to lengthen or widen their penises (the procedure has also been performed on men in Japan, Australia, Germany, Britain, and other countries). The lengthening procedure does not involve any plastic inserts or fleshy attachments—the surgeon cuts two ligaments that attach the penis to the pubic bone and this allows the penis to extend fully from the body, externalizing a portion usually hidden inside the body. As one plastic surgeon said, "penis exposure" might be the most appropriate term for the procedure. After the penile suspensory ligaments are released, the angle of erection is less acute but the penis gains about one to two inches in length. Penises are widened by the injection of fat, although this procedure is not always aesthetically successful since the head or glans cannot be enlarged. This may leave the organ looking a bit like a tied balloon.

Why do men bother? One physician studied three hundred men

(of one thousand applicants in one clinic) who were approved for penis-enhancing surgery. The majority expressed what the physician, Dr. R. H. Stubbs, called "locker room phobia," a sense of discomfort in front of other men. Fewer than a third came as a result of complaints or criticisms from their female sexual partners. The locker room phobia suggests that males view their penises as a means of competing with other males.

In many primates, males present their genitals as a threat and display of dominance. Ethologist Irenaus Eibl-Eibesfeldt has observed that, when groups of vervet monkeys feed, several males sit with their backs to the group and display their vivid red and blue genitals. If an unknown animal approaches, they get an erection and make a threatening face. The erection is an action manqué—a ritualized threat to mount. People in various Asian and African cultures carve figures that display threat faces and erect penises as guardian figures to protect the home and to wear as amulets. Although a skinny human with a large penis is not more threatening than a muscled hunk with average or even small genitals, a large penis may still act as a threat display (think of the impact of "flashers" and compare that with the exposure of any other body part by either sex).

Perhaps penises are not (only) weapons of war but charms to lure females, and in that case the larger the better as animal attraction would show us. In most societies today, women do not get to see the penis immediately, so its value as a courtship device is questionable. By the time most women see what they're getting, they've already been successfully courted. But we didn't always wear clothes, and it's useful to think about the aesthetics of genitals in the environment from which we evolved.

To Freud, such a consideration was impossible: "We never regard the genitals themselves, which produce the strongest sexual excitation, as really beautiful." Whether or not Freud is right, women, whether out of modesty or lack of inspiration, have not penned paeans to the penis. Sylvia Plath's famous glimpse of her lover's genitals in *The Bell Jar* is indelible: "turkey neck and turkey gizzards." Camille Paglia warns that penises risk "ludicrousness by their rubbery indecisiveness," but have a "rational mathematical design, a syntax" that female genitals lack (she is even less impressed with the female genitals, which she describes as "vagrant in contour and architecturally incoherent"). Male centerfolds have never caught on with a female audience.

But visuals are not everything, and most studies of women suggest that in sexual encounters tactile cues are as important as visual cues. William Eberhard, a professor of biology at the University of Costa Rica, believes that female selection pressure has shaped male genitals in a variety of species to be "internal courtship devices" whose beauty might rest more in what they do than in the way they look. Although the standard thinking in sex research since Masters and Johnson in the 1960s is that size does not matter, there is little evidence one way or another. Pornography certainly emphasizes huge phalluses and, by implication, that size matters at least in terms of the man's self-esteem and the partner's initial impression.

What the size of the male genitals does tell us is quite a lot about the sex habits and preferences of the females they mate. Let's first look at testicles. Gorilla males guard their female harems and face little outside competition. The gorilla has an enormous body and a tiny penis and testes. Chimpanzees, on the other hand, live in a promiscuous society with unstable dominance patterns. The males are only slightly larger than the females but they have enormous testes. Their big balls are not just ornamental. They manufacture more sperm per day and introduce more sperm at each intercourse than smaller ones. If several males are inseminating the same female, the male with the most sperm will have an edge. Human males have testes proportionately larger than the gorilla's but proportionately smaller than the chimp's, suggesting that our ancestral females were not as promiscuous as chimps, but not entirely monogamous either (if human males had not faced any sperm competition they would most likely have developed tiny testes like the gorilla).

Although human males have the proportionately largest penises of the primates, William Eberhard has found what he calls "genitalic extravagance" in male animal species from nematodes and flies to the human. Many male genitals are more elaborate and complex than those of the female, and more elaborate than they need to be to simply deposit sperm. As entomologist James Lloyd notes, penises can come equipped with "little openers, snippers, levers and syringes. . . . collectively a veritable Swiss army knife of gadgetry." The genitalia of many insects are so elaborate and so dissimilar in shape that they provide the most reliable means of identifying the species.

Some scientists believe that genitalia evolved into elaborate shapes as means to prevent hybridization. The male and female genitals of each species fit like lock and key, because only members of the same species have matching equipment. Others have noted that male genitals are at their most elaborate in species in which the female mates with more than one male, suggesting that competition among potential mates may be driving the genitalic extravagance. Animal ecologist Goran Arnqvist pitted these two ideas about the origin of genital extravagance—the lock and key idea versus sexual selection—against one another by studying insect females that mate only once and females that have multiple mates. If the lock and key idea is correct, the males that mate with the monogamous females should have the most extravagant genitalia since the cost of mating with an inappropriate male would be enormous for the female. Instead, he found that the male genitals were far more varied in the group whose females had several mates competing for her.

It appears that some male genitalia may have been shaped by female taste. It is possible that the large human penis was selected because it titillates or stimulates the female and thereby promotes conception, or because it is better at dislodging the sperm of other males and placing its own more advantageously.

Even Handed

When it comes to pleasing the female, however, other researchers say to look not only at size but at symmetry. Behavioral ecologist Randy Thornhill and psychologist Steve Gangestad say that a man who develops more symmetrically—that is, a man whose feet, ankles, elbows, hands, wrists, and ears are equal in width—is more attractive to women and better in the bedroom (though he may be a cheating louse) than a man who is less evenly matched. Remember that, in animals, symmetry is a sign of good development, parasite resistance, survival, and fecundity, not necessarily loyalty, fealty, or devotion.

It is unlikely that anyone notices subtle differences in wrist size or that any woman is turned on by symmetrical ankles, but men with symmetrical bodies tend to have other attractive features, such as symmetrical faces (which, as we have seen, are considered attractive), and bodies that are more muscular, taller, and heavier than those of men with less symmetrical bodies. The Japanese scorpionfly female prefers a male with symmetrical forewings. What lures her? His pheromones: symmetrical males attract females with their chemical signal even when the female cannot see the symmetrical male. Bees are attracted to symmetrical flowers because they turn out to have more nectar. Symmetry is the hidden persuader, correlated with attractive scents, nectars, and faces.

Symmetry is tied to beauty because it acts as a measure of overall fitness. Stressors, including inbreeding, parasites, and exposure to radiation, pollutants, extreme temperatures, or marginal habitats can interfere with the precise expression of developmental design during the growth of symmetrical traits such as horns, antlers, petals, tails, wings, ankles, feet, faces, or whole bodies. Fluctuating asymmetry may be the body's most sensitive indicator of its ability to cope with stress during development. Symmetrical animals have higher growth rates, are more fecund, and survive longer.

In a review of sixty-two studies conducted with forty-one species, zoologist Anders Moller and behavioral ecologist Randy Thornhill found that fluctuating asymmetry was associated with mating success or sexual attractiveness in seventy-eight percent of species, including the human species.

Men with symmetrical bodies report that they start having sex three or four years earlier than other men, have sex earlier in the courtship, and have two to three times as many partners. They also are more pleasing to their lovers. Thornhill and Gangestad studied the sex lives of eighty-six heterosexual couples in their twenties. Women with symmetrical partners (as measured by calipers placed on their elbows, feet, and so on) had significantly more orgasms during intercourse, as reported by both partners.

Fluctuating asymmetry turned out to be a better predictor of female orgasm than the couple's feeling of love, the investment of either party in the relationship, the male's potential earnings, or the level of sexual experience or frequency of lovemaking of the couple. When it came to orgasm frequency and timing, physical allure ruled. Symmetrical men had bigger, taller, and more muscular bodies as well as more attractive faces, and these also were related to the probability of their partner's experiencing orgasm. But symmetry still made its own independent contribution. If we accept the notion that female orgasm is a form of female choice, a way that she influences reproductive outcome, it seems that females favor the hunks—the men with big, symmetrical bodies and handsome faces. But they are also favoring the cads. As it turns out, symmetrical men are more unfaithful and invest less in their relationships than asymmetrical men. Lest we think that women favor cads per se, there may be a simpler explanation. The symmetrical men are favored by women and may be receiving more competing offers than other men.

Symmetrical women are favored too. They have more sexual partners than less symmetrical females, and may be more fertile. One study found that women with large and symmetrical breasts were more fertile than women with less evenly matched breasts. Interestingly, women's symmetry changes across the menstrual cycle. They are most symmetrical (and presumably most attractive to their partners) on the day of ovulation. Biologist John Manning measured the size of the left and right ears, and the third, fourth, and fifth fingers of each hand in thirty healthy women between the ages of nineteen and forty-four. The timing of their ovulation was confirmed by a pelvic ultrasound. He found that asymmetries decreased thirty percent in the twenty-four hours prior to ovulation.

Bust Lust

In 1947, John Steinbeck wrote, "A visitor of another species might judge . . . that the seat of procreation lay in the mammaries," so preoccupied are men with women's breasts. Anthropologists claim that there are cultures in which breasts are not considered sexy. But it is unclear if there is any place on earth where men don't find the breasts of healthy nubile girls beautiful.

Female breasts are like no others in the mammalian world. Human females are the only mammals who develop rounded breasts at puberty and keep them whether or not they are producing milk. Other mammals have breasts that swell only when they are full of milk, collapsing when breast feeding is over. Breasts are not sex symbols to other mammals, anything but, since they indicate a pregnant or lactating and infertile female. To chimps, gorillas, and orangutans, breasts are sexual turnoffs.

In humans, breast size is not related to the amount or quality of milk that the breast produces. Anthropologist Bobbi Low has suggested that breasts evolved as deceptive signals, to give the female the appearance of having a good supply of future nutrients for offspring. But others argue that the breast as milk jug and the breast as tinder for lust are not one and the same. Men will see breasts as sexier the less they remind them of feeding stations. Others suggest that breasts honestly advertise the presence of the fat reserves needed to sustain a pregnancy.

There is some cost to having permanent breasts, as opposed to the collapsible models of the chimp. As any female athlete knows, the larger the breasts the more they interfere with running and jumping, and moving the arms and the upper body. The mythological warriors of ancient Greece, the Amazons, removed their right breasts because they got in the way of their archery. Today, there is a thriving business in sports bras aimed at minimizing bounce and maximizing protection of this injury-prone area. One of Japan's largest producers of sports goods has recently created a swimsuit that borrows from aircraft and maritime technology to decrease the drag that results from having breasts.

What is their benefit? Well, for one, men think they are beautiful, and so do many women. Desmond Morris suggested that humans developed large rounded breasts as a way to shift the male interest toward the front and encourage face-to-face bonding. Most animals have brightly colored and fleshy rumps, and they mate from behind. As humans developed face-to-face positions, the body evolved attractors where men's eyes are. Morris says that we find large, upright breasts appealing because they mimic the appearance of sexual excitement, when breasts get firmer and more rounded and the nipples pointier.

Whether or not breasts evolved to keep men and women face to face during sex instead of glancing behind, clearly their beauty depends on shape and changes with time. The erotic breast has always been firm and upward tilting, no matter what the preferred size. It is never flabby, elongated, or tubular in shape (the latter mimicking an older breast even if present in a young girl). From the paintings of the Renaissance, when rounded breasts were placed high on the chest "beyond the laws of gravity," to the torpedoes launched by the 1949 Maidenform bras, to the raised globes produced by corsets, Wonderbras, and plastic surgery, the ideal breast has the shape found naturally only in young nulliparous females.

It is likely that breasts were once considered sexy only in young women. Once they became used regularly for breast feeding, and changed shape, their value as erotic objects was lessened. As Donald Symons noted, "After a few years of nursing, breast shape probably actually hurt attractiveness, because it is an unambiguous cue of age and parity." As an anthropologist once explained to me, cultures around the world make three distinctions among breasts, which he designated by putting his palms flat against his chest, pointing his hands straight out from his chest, and pointing his hands at a downward angle from his chest. The first indicates an infertile young girl, the second a nubile girl, and the third a woman who has borne and fed a child, and/or gotten older.

Female décolletage first became a fixed style during the Renaissance at a time when women of the upper classes hired wet nurses to feed their babies. As Marilyn Yalom points out, ninety percent of the population at this time were milk bearers, and only ten percent were not. Having a wet nurse was a status symbol and breast feeding was associated with lack of money. Portraits of the elite showed them with high round breasts, in contrast to the pendulous breasts of peasants or wet nurses. The upper-class clothing styles squeezed the breasts up above tight corsets that cinched in the waist. Tight sleeves were worn to slim the arms. From this constricted upper torso emerged two rounded, even breasts. Décolletage has remained a part of elite formal wear ever since.

During the years 1940 to 1970, breast feeding declined in the United States owing to milk substitutes and heavy promotion of bottle feeding. At the same time, there developed a renewed interest in displaying the breast as an erotic object. In 1943, Jane Russell became one of the first mammary queens when she showed off her breasts in the movie *The Outlaw*, wearing a metallic uplift bra designed by her admirer, Howard Hughes. In the 1950s and 1960s actresses with large breasts became screen idols—Lana Turner, Diana Dors, Marilyn Monroe, Brigitte Bardot, Gina Lollabrigida, and Jayne Mansfield.

Not satisfied with padding or torpedo bras or bust developers, women turned to permanent body alterations when they became available. By 1991, over two million women had had breast implants. In the breast implant heyday, 120,000 to 150,000 were being done a year. In February 1992 the FDA severely limited the use of implants because of concerns about their safety. But by 1998 the numbers have climbed back up to over 122,000 operations. Silicone gel is still banned for cosmetic use, but solid silicone forms filled with saline are not. The latest trend is for even larger implants than ever.

It is too early to know the exact dangers of silicone implants but, no matter what they turn out to be, we know that it is a risky business to implant a foreign substance into disease-prone regions. Pain, numbness, bruising, and the formation of tough fibrous capsules around the implants are not uncommon, nor are rupture and leakage. Certain breast surgeries may also interfere with mammography and breast feeding. Another hotly contested question is whether silicone migration causes damage to the immune system. The saline implants are considered safer, but the other problems—infections, scar tissue, and rupturing and interference with mammography—remain.

In the United States breasts have become a major focus of erotic interest, an interest that has coincided with lower incidence of breast feeding and with increased use of surgical techniques to mimic the nulliparous breast. It is interesting that in our lipophobic society, where fat has been demonized practically everywhere else on the body, there is still so much interest in acquiring and admiring these two rounded vessels plumped with fat.

Waist Land

But if fat appears a few inches south, out come the ab machines, the girdles, the liposuction, and the diet books. So welcome in breasts, fat is loathed when it appears on the waist. Wasp waists are sexy—not as obviously sexy as breasts but they are another one of those hidden persuaders, like symmetry, that have a profound impact on the way bodies are viewed.

The waist is a zone of organs and muscles located between the rib cage and the iliac crest of the hips, whose shape is defined by fat and muscle (and by the health of internal organs). Healthy premeno-pausal women have waist-to-hip (WHR) ratios of .67 to .80. This means that their waists are seven to eight tenths as large as their hips. A healthy man's waist-to-hip ratio tends to be between .85 and .95. Testosterone stimulates him to deposit fat in the abdominal regions (and the nape of the neck and the shoulders) and inhibits him from accumulating it in his hips and thighs. This gives him the male "android" body shape—a tendency to develop belly fat. The female or "gynoid" shape is the hourglass, where fat accumulates around the hips. Moderate weight gain does not alter these basic male and female shapes, and they are found all over the world among people who vary considerably in height and weight.

The female or "gynoid" shape emerges at puberty under the influence of estrogen. Young women gain body fat and store more of it in their thighs and buttocks than anywhere else on their body. In fact, thighs account for one quarter of women's weight, which is why in 1996 thigh creams were a ninety-million-dollar business, although there is no evidence that they really work. But women are desperate. Diets will tend to take weight off the upper body and the breasts before the thighs or buttocks. This is because fat in these regions is rarely used by the body except during pregnancy and lactation. It appears to be deposited there for this reason, insuring that the body has enough stored calories to successfully complete a pregnancy and lactation even during an ensuing famine.

The small female waist, poised between the rounded breasts and hips, has an ephemeral beauty. It disappears early in pregnancy and is hard to regain after pregnancy. By menopause many women have a waist-to-hip ratio that is closer to a man's than to a younger woman's. The waist is one of the body's best indicators of hormonal function. Women with polycystic ovary disease, a condition attended by elevated levels of testosterone, have masculine waist-to-hip ratios. Too many androgens and the body starts to accumulate fat in the abdomen rather than the hips.

Two large-scale, well-controlled studies of fertility convincingly link waist-to-hip ratio and women's reproductive potential. In a study of five hundred women who came for artificial insemination to a clinic in the Netherlands, fat distribution actually made more of an impact than age or obesity on the probability of a woman's conceiving. A woman with a waist-to-hip ratio below .8 (small waist and an hourglass shape) had almost twice as great a chance of becoming pregnant as did a woman whose waist-to-hip ratio was above .8 (thicker waist and more tubular shape). In another study of women attempting to conceive through in vitro fertilization and embryo transfer, a waist-to-hip ratio above .8 was again negatively associated with chance of pregnancy. The impact remained even after the authors took into account the women's age, body mass index, and history of smoking.

If men are looking for fertile mates, it's no coincidence that narrow waists should look attractive to them. Psychologist Devendra Singh has tested men's perceptions of body shape in eighteen cultures. He finds that waist-to-hip ratio is often more important than breast size or weight (barring extremes) in making a woman's body appear attractive to men. He shows people line drawings of women at three different weights (underweight, average, and overweight) and three different waist-to-hip ratios (.7, .8, and .9) and asks them to choose which figure is the most attractive. Overwhelmingly, men in his studies have chosen the average-weight woman with the .7 waist-to-hip ratio as the most attractive.

Dev Singh believes that men have an innate preference for female bodies with narrow waists and full hips, which signal high fertility, high estrogen, and low testosterone. As in all things, a little exaggeration is sometimes welcome. Singh finds that figures with .6 waist-to-hip ratio are also considered attractive. Barbie, another example of a sex bomb, comes in at 36-18-33, with a .54 waist-to-hip ratio.

If one looks at icons of beauty, Singh's point about the importance of body shape becomes apparent. Audrey Hepburn and Marilyn Monroe represented two very different images of beauty to filmgoers in the 1950s. Yet the 36-24-34 Marilyn and the 31.5-22-

31 Audrey both had versions of the hourglass shape and waist-to-hip ratios of .70. While some have claimed that Americans are starting to favor tubular boyish bodies, Singh says that this is not true. Looking at Miss Americas from the 1920s through the 1980s and at *Playboy* from 1955 to 1965 and 1976 to 1990, he found Miss Americas' waist-to-hip ratios varied only within the .72 to .69 mark, and *Playboy* models within the .71 to .68 range.

The current average supermodel measures 33-23-33, which gives her a .7 waist-to-hip ratio. Elle MacPherson, known as "The Body," has not only 44-inch legs, but measurements of 36-24-35, in other words a curvy .69. The models get taller, the breasts get bigger, the hips may whittle, but the gynoid shape remains, because healthy young females deposit more fat on their hips than their abdomens.

Psychologist Martin Tovee and his colleagues in England came to the same conclusions when they studied the vital statistics of three hundred supermodels (including Claudia Schiffer and Naomi Campbell), three hundred "glamour models" from *Playboy*, three hundred average women, and smaller groups of women with eating disorders (both anorexics and bulimics). The supermodels were considerably taller than all other groups, and both supermodels and glamour models were thinner than the average women (though considerably heavier than anorexics). Both supermodels and glamour models were curvaceous, with waist-to-hip ratios of .71 for the fashion models and .68 for the glamour models. Glamour models may appear even curvier because they are shorter than supermodels, but even fashion models are not what Tovee calls "stick insects."

Although models have been blamed for the rise in eating disorders in young women, Tovee found the body shapes of the two groups very different—suggesting that no amount of dieting can bring about a model's shape. Models are statistical rarities who can combine tall lean bodies with curves. Indeed, just remember Twiggy. In her modeling heyday she was ninety-two pounds and 31-24-33, giving this supposed androgene a waist-to-hip ratio of .73—even Twiggy deposited some hip fat.

Aside from advertising reproductive potential, a slim waist is generally a good indicator of good health. Belly fat in both sexes is associated with greater risk of heart attack, diabetes, stroke, hypertension, gall bladder disease, and some cancers. If one is looking at health alone, a slim waist would be desirable. However, we must remember t'iat these chronic diseases are largely products of modern life. Thus, although waist-to-hip ratio is an excellent modern indicator of health in general, its primary evolutionary importance for detection of beauty probably had more to do with what it signified about fertility.

In the United States obesity has such a great influence on people's perceptions of attractiveness that neither breast size nor waist-to-hip ratio can compete. While men prefer the hourglass shape, it won't lead them to prefer an obese woman with a waist or large breasts over a thin or average-weight woman with a more tubular body. In Devendra Singh's studies the heaviest figures are judged to be eight to ten years older than the slimmer ones (even though the faces are identical). Their decreased youthfulness may contribute to their lessened attractiveness. But in our fat-phobic culture issues of social status are likely to play their role too, and social status and body fat have a significant negative correlation, particularly for women.

Women's fashion has consistently called attention to the waist with corsets, wide belts, hip huggers, and crop tops. Corsets have been a fashion constant for five hundred years. According to the Oxford English Dictionary, the first use of the word "corset" was in a 1299 account of the wardrobe of the household of King Edward I. Not only were waists thinned and busts pushed higher by corsets, but hips were fanned out with pads, cushions, and wire cages called farthingales, panniers, crinolines, and bustles. Some widened the woman at the side, some encased her in a cone, others plumped up her behind. Women escaped these caricatures of the female shape only early in the twentieth century when designers Madeleine Vionnet and Paul Poiret introduced more fluid styles. But fashion continually comes back to highlight the waist, from Dior's New Look of 1947 with cinched-in waists and a wide skirt to playful renditions of corsets by Vivienne Westwood and John Paul Gaultier.

The desire to exaggerate the female hourglass may help to explain a seemingly unrelated fashion phenomenon—high heels. One would think that, if anyone wanted heels, it would be men, since they are much more obsessed with their height. Platform shoes were originally worn by both sexes but as heels evolved from chunky in shape to long thin spikes they were meant exclusively for women. Heels reached their apogee in 1953 with the stiletto, created in Italy and popularized in Paris.

Heels are popular with women because they do much more than add inches. Model Veronica Webb put it bluntly: wearing heels is "like putting your ass on a pedestal." Balancing precariously on the balls of their feet, wearing heels forces women to throw back their shoulders and arch their backs, making their breasts look bigger, their stomachs flatter, and their buttocks more rounded and thrust out. And this is just an aside from what they do to the shape of the leg, which appears more toned, elongated, and recalling the shape of a leg tensed by arousal.

And then there is the walk. Some have said that the high-heeled walk is sexy to men because it hobbles the female—she looks fragile, and unable to get away quickly. Others, however, think it's the way the hips move. As premier shoe designer Manolo Blahnik said, "I've been in this business for twenty-three years, I've done platforms, I've done wedges—nothing gives you that feminine walk like a pair of high heels. I wouldn't be here if women could resist heels."

Critical Mass

In the United States and in much of the Western world, to be beautiful means to be lean and lithe, as sylphlike as a ballerina and as self-denying. In the midst of food abundance and culinary temptations, our highest-paid fashion models are five-feet-nine-inch women who weigh one hundred and ten pounds. Peering at them enviously is the average five-foot-four-inch, one-hundred-forty-two-pound woman.

Weight is often calculated as the Body Mass Index, which is weight in kilograms (pounds multiplied by .45) divided by the square of height in meters (inches multiplied by .0254). The United States government recommends that people maintain BMIs under 25. For a woman five feet four inches this means weighing under one hundred and forty-five pounds. Healthy as she may be at this weight, her body will be unfashionably large unless she whittles off another thirty pounds.

But she will have a hard time doing so. The average American keeps getting heavier despite a collective pouring of forty billion dollars a year into weight loss centers, health clubs, diet meals, low-calorie sodas, appetite suppressants, exercise equipment, and exercise videos. One third of the population is obese (defined as twenty percent over accepted standards)—this means thirty-two million women and twenty-six million men in the United States. The figure jumped between 1980 and 1991 after it had held steady at twenty-five percent for three decades. Weight woes are reported in Britain, where the proportion of obese adults has risen to thirteen percent of men and sixteen percent of women, double what it was ten years ago, and in many other developed countries.

If we are as influenced by media images as many social critics claim, one would think that most men and women would be lean or at least getting leaner by now. The huge amount of money poured into the slimming industry suggests that we are trying . . . or are we? Fat consumption in the United States dropped only two percent in the 1980s (from thirty-six percent to thirty-four percent of caloric intake) and high-fat french fries, hamburgers, and chicken nuggets were three of the five fastest-growing restaurant foods. Portions have gotten so large at restaurants that plates have gone from ten inches to eleven or twelve to accommodate the heaping portions. We aren't doing any better using those exercise videos or treadmills either. The average American spends only sixteen minutes a day exercising. Twenty-four percent of Americans are totally sedentary couch potatoes and tomatoes. A person who spent all day in bed would need between 1,400 and 1,600 calories to properly maintain the body and all its silent operations (the replacement of the cells that line the intestines, bladder, and so on). Active people need no more than 2,000 calories. With so much food available, and so little energy expended, it's no wonder that waistlines are expanding.

We are not lazy or hypocritical, gluttonous or evil, we are human. And we are up against millions of years of evolution that have selected for the ability to eat heartily, store fat, and take in as much fat, salt, and sugar as we can. We are adapted to a world of periodic famines caused by droughts, floods, earthquakes, and the scarcity of plants and game. Even today, almost half of developing societies have food shortages at least once a year, and a third of those have shortages that are severe. That the body has a propensity to store fat, and to respond to food shortages by resetting the metabolism and using food more efficiently, is the bane of dieters but highly adaptive.

Or at least it was adaptive before we raised grain-fed animals in penned corrals that yield prime sirloin with thirty percent fat, or refined sugar to create éclairs and doughnuts. As with all our other pleasures, we tend to go for exaggeration, not subtlety. Every large supermarket and pharmacy has aisles devoted to candy bars, and every summer the barbecues fire up and the air smells of dripping fat.

Until recently we could afford to eat what we wanted because the food supply was limited in both quantity and variety (as anthropologist Napoleon Chagnon has said, hearts of palm may be a delicacy in the United States, but when he lived with the tribal Yanomamo and had only hearts of palm to eat at times, they lost their appeal). And until this century our output matched our input. We had to be active because there were no cars to transport us, supermarkets to provide us with packaged foods, gadgets to do our household chores, and stores of every variety to provide for our every need. In the world in which we evolved, it made sense to hoard rest time. Today we may spend the day sitting on an office chair, then sitting in the car, and finally spend the night sitting on the couch watching television, all the while trying to force ourselves to get on the treadmill or go to the health club.

Why we became obese is not mysterious; we have plentiful food, bodies not equipped with sufficient brakes for fats and sweets, and we have arranged the world so we need less and less physical exertion to survive. We are forced to do the unnatural: refuse food and engage in purposeless activity for the sake of burning it off and keeping our bodies tuned up. What took centuries to develop in other cultures happened quickly in islands like the Tongan cluster in the South Pacific. A 1986 survey found that two thirds of Tongan women in their forties were obese; diabetes, hypertension, and heart disease were epidemic and life expectancy had decreased. Rapid changes in lifestyle caused the epidemic of obesity. Cars were introduced, and foods were imported for people who had eaten yams, manioc, and fish that they had either planted or caught. Without knowledge of health disadvantages or an aesthetic of thinness to rein in the appetite, many happy Tongans were eating whole loaves of bread spread with half a pound of butter, and ice cream. Mutton flaps too fat and greasy to be considered edible by New Zealanders were shipped over and became a favorite fried daily staple. The World Health Organization needed to convince their leader King Taufa'ahau Tupou IV of the importance of weight loss. Like many leaders, the king was a big man—he was six feet two and weighed 441 pounds. In 1976 he had been listed as the world's heaviest ruler. Convinced that he needed to restore the reputation of Tongans as strong athletes and not fatties, the king slimmed down and had his gym sessions videotaped for the masses. They now join many Westerners in dieting and yet remain overweight.

It is said that non-Western cultures have a beauty ideal of fatness. Ancient fertility goddesses such as the Venus of Willendorf show massive breasts, hips, and torsos. Among the Annang of southeastern

Nigeria and in other west African countries, girls are housed in special "fattening rooms" where they are secluded and fed to make them plumper in preparation for mating. The extra body fat helps bring on menstruation and fertility and increases the chances of a healthy baby (heavier mothers produce heavier babies, who grow and develop faster). The weight gives the girl and the family social status, because it demonstrates that the family could afford to siphon off resources to make their daughter plump and fertile. However, few girls can afford to stay in long enough to get "fat" and whatever fat they do accumulate is transient. Evolution would have left us with beauty detectors attuned to the normal or average body, which was lean.

And in fact this is what most studies find. The normal or average weight is usually preferred to the very slim or heavy. Devendra Singh finds this in the eighteen cultures he has tested, and psychiatrist Anne Becker has found this even in Fiji, another South Pacific island with an extremely high prevalence of overweight and obese people. Unlike in the United States, however, the two thirds of both males and females who are overweight or obese do not show an overwhelming desire to lose weight nor do they torture themselves over a slim ideal. Over fifty percent of the obese females she interviewed and more than seventy percent of the overweight females wanted to maintain their current weight. But shown line drawings of thirteen figures ranging from very thin to very obese and asked to pick the most attractive, the Fijians selected figures at the mid-range. Very similar figures have been shown to people in Britain, Uganda, and Kenya by psychologist Adrian Furnham and his colleagues. There is striking cross-cultural consistency in which figures were found most attractive. None rated the extremely fat or thin attractive, and all picked the same cluster of mid-weight figures for both male and female drawings. Anne Becker found that, within Fiji, the ratings were consistent across both sexes, and no matter how close or far a village was from an urban center.

Then what would cause certain societies to have an ideal of

weight that was so far from the average? The Darwinian forces are multiple and sometimes contradictory—feeding yourself is one, and attracting mates is another, and the mate attraction motive is tied both to health and to status. The status motive can sometimes lead to impulses that counteract the feeding motive and override our aesthetics. We may know that skeletal or obese bodies don't look as good as average-weight bodies, but we value them as emblems of status and desire them the same way we do any other accoutrements of wealth. Anthropologist Margaret MacKenzie writes, "In a context where only a King can control enough food resources and labor supply to eat enough and do no physical labor so that he becomes fat," prestige is conferred by signs of abundance. A thin person is a person too poor to afford the calories, and maybe one who does so much physical labor that she cannot keep weight on. When poor women are fat (because junk food is so cheap and available, and they are less educated about its hazards and unable to afford expensive healthy foods), then it's in to be thin and dietary restraint and physical exercise become prestigious.

Obesity researchers Jeffrey Sobal and Albert Stunkard reviewed one hundred and forty-four studies of the relation between socioeconomic status and weight. They found a strong inverse correlation between a woman's weight and her social and economic status (the higher the status, the lower the weight) in Belgium, Britain, Canada, Czechoslovakia, Germany, Holland, Israel, New Zealand, Norway, Sweden, and the United States and virtually all developed countries. They found a relation every bit as strong in the opposite direction in developing countries with food scarcities. There the higher-status men and women were heavier. In the developed countries, the relation between status and weight was less consistent for men.

The thin ideal is maintained by the high in status through diet and exercise. It is also maintained by social mobility—thin women are more likely to "marry up," to marry men who have higher social and economic status than their family of origin. Studies in the United States, Germany, and Britain all find that upwardly mobile women are much thinner than their counterparts who marry men of the same social class or lower. There is also a genetic component. When identical twins overeat, they gain almost identical amounts of weight and tend to store it in the same regions. Like height, weight is determined by many factors and the genes set limits to how much we can vary the shape of our bodies. But the acquisition is determined by life conditions.

The most extreme manifestation of thinness came from the world of high fashion; in fact it trickled down from this world to the masses over the course of the century. Two British girls, Twiggy in the 1960s and Kate Moss in the 1990s, one five feet six and the other five feet seven and each skimming under one hundred pounds, became the thin ideal's poster children. As James Wolcott wrote in *The New Yorker*, Kate Moss became a "dartboard for every point to be made about bad female body image." These images influenced mass media representations of beauty. The Miss Americas of the 1960s were five feet six inches tall and weighed a hundred and twenty pounds. Twenty years later they had gained two inches in height but remained at the same weight. *Playboy* playmates also dropped several pounds and gained several inches in height during the same period, dipping from eleven percent below the national average to seventeen percent below it. Even the Columbia Pictures logo, the torch-bearing woman, was slimmed in 1992.

At the same time that the thin ideal has become entrenched, it has been vilified by feminists and cultural critics as oppressive and dangerous to women. Critics say that models project the wrong image of women, the look of forestalled maturity and fragility. Some say they look like heroin addicts, or that the ads smack of child pornography. The most often repeated criticism suggests that their waiflike bodies are responsible for spreading a contagious disease—that they have created an epidemic of eating disorders among teens. The first alarm bell was rung in 1978 by Hilde Bruch when she described anorexia as "an epidemic illness." She caught the world's attention with these lines: "New diseases are rare, and a disease that selectively befalls the young, rich and beautiful is practically unheard of. But such a disease is affecting the daughters of well-to-do, educated, and successful families. . . . for the last fifteen or twenty years anorexia nervosa [has been] occurring at a rapidly increasing rate. Formerly it was exceedingly rare." A few years later, *Newsweek* pronounced 1981 "the year of the binge purge syndrome," and thus brought a second eating disorder, bulimia, to the attention of the public.

Naomi Wolf probably scared every mother in America when she wrote that "it is dead easy to become an anorexic," suggesting that every dieting teen was susceptible. Women, she said, "must claim anorexia as political damage done to us by a social order that considers our destruction insignificant because of what we are—less." Such words valorized the anorexic as a symbol of the struggles of contemporary women to inhabit their bodies. Wolf's book struck a chord with young women who were tired of battling their bodies, and tired of being unable to meet a beauty ideal or to profit by their own measure of it.

How prevalent are eating disorders and what causes them? Dieting and overeating have become so ubiquitous in our culture, it's hard to even define normative eating, but as historian Joan Jacobs Brumberg has pointed out, eating disorders did not arise *de novo* in the 1960s. In terms of the general population, they are relatively infrequent, and even among young women they are unusual. The prevalence of bulimia is one to five percent of college-age women in the United States. Anorexia is a rarer disorder that tends to develop in late adolescence, before age twenty, and strikes a half to one percent of girls. Anorexia is the more dangerous disorder, and there is the small but not insignificant chance of death since the anorexic may restrict herself to 200 to 400 calories a day. The bulimic alternates between food restriction and binges when she may eat 8,000 calories at a time. Ninety to ninety-five percent of the sufferers are women, and the vast majority of these are white and from the upper social class. But the disease has a "social address," not a racial or sexual restriction. Gay men are showing increasing incidences of eating disorders, and as women of color move into the upper classes, their likelihood of developing an eating disorder rises.

Although it is tempting to blame advertising and fashion for eating disorders, one must keep in mind that ninety-eight percent of women do not develop diagnosable levels of eating disorders although all are exposed to the media. Most women diet, many develop extreme dissatisfaction with their bodies, but very few drop to under seventy pounds and stop menstruating, or eat four days' worth of high-fat food in one sitting. It is not "dead easy" to develop anorexia or bulimia; they are complex and serious disorders with multiple etiologies that afflict particularly vulnerable young women and some young men.

There is no evolutionary precedent for the slim ideal. Matter of fact, selection should work against such a preference. It has been known for some time that women with eating disorders suffer disruptions in fertility and reproduction. Even rats and mice placed on severely restricted diets (but given nutritionally sufficient amino acids, vitamins, and so on) suppress estrous cycles and decrease in fertility. Starved animals don't reproduce, they don't even mate. Closing down fertility is the body's adaptive safety valve, allowing it to avoid having to satisfy the high food demands of pregnancy during times of famine. Recently, however, scientists have found that these food-restricted animals live longer (up to thirty percent longer). They live longer and their fertility remains in a state of suspended animation awaiting the food supply that will catapult them into reproductive maturation. The pause is real—middle-aged mice that have been food-restricted have ovaries that age more slowly.

If that sounds as though it may have interesting implications for the modern woman, the idea has occurred to some psychologists. They suggest that restricted eating by women may be a nonconscious strategy to control reproduction. Eating disorders and the diet craze became prominent in the 1960s when women gained relative sexual freedom and financial independence, and wanted to delay reproduction. Another way of looking at the phenomenon is to consider that women, not men, are driving the thin ideal, and it may help to explain the idealization of thinness among the highest social classes.

But if extreme slimness suggests lack of fertility, men should be staying away in droves. Well, not when women provide all the right contradictory cues—tall, symmetrical bodies with large breasts (which should be impossible at that weight) and curved hips. Since thinness has also become a status symbol, this adds to its allure to men, who want to be with a woman that other men and women find attractive.

Extreme thinness is a fashion, a fashion set by the highest social classes, as most fashions are. Our bodies reflect not only Darwinian forces which impel us to reproduce, but cultural ones, and social ones, and these are most brilliantly displayed in fashion.

CHAPTER SEVEN

Fashion Runaway

It is clear that between what a man calls me and what he simply calls mine, the line is difficult to draw.

—WILLIAM JAMES

You may have three-halfpence in your pocket and not a prospect in the world . . . but in your new clothes you can stand on the street corner, indulging in a private daydream of yourself as Clark Gable or Greta Garbo.

—GEORGE ORWELL

No one is able to resist that delicious itch to reveal his own picture of himself through fashion.

—TOM WOLFE

Fashion dies very young, so we must forgive it everything.

—JEAN COCTEAU

At the Paris couture shows a Thierry Mugler dress hangs from a model's nipple rings and a John Galliano dress trails for twelve feet. In Burma a woman of the Padaung tribe wears a ten-pound coil of brass rings around her neck, and coils of metal around her calves and wrists. On the streets of New York City girls layer miniskirts over pants and walk on platform shoes. It is unlikely that any one human could find all of these adornments beautiful, yet someone always does, somewhere, at some time.

To the critic who says that taste in fashion is not universal, I say: you are right. Ovid said, "I cannot keep track of all the vagaries of fashion. Every day, so it seems, brings in a different style." He was talking about hairstyles, but it is true of fashions generally. A flip through the history books will show that a universally admired adornment for the head or foot or for any body part in between does not exist.

Fashion can help illuminate what we find beautiful and why (in what it exaggerates in the human form, and what it attempts to cover up) but it is different from beauty itself. The seismic shifts in fashion, though striking, have little to do with beauty. Although a chapter cannot do justice to a topic as rich as fashion, I include one here to point out the many things that human beauty is not, but is often confused with.

Quentin Bell likened fashions to fruit flies—short-lived species whose quick mutations make the former an ideal subject for the sociologist and the latter for the geneticist. Fashion is an art form, a status marker, and a display of attitude. We create it, as we create architecture and furniture, to help us negotiate our relations with the outside world and to provide us with comfort and protection. But as visual extensions of our persons, they also mirror our desires in complex ways.

Fashion is of the moment. The most expensive clothing makes its debut in a setting of live theater: months of work culminate in a brief, dazzling display before an invited audience. Great clothes are all about making a drop-dead entrance. Contemporary fashions may allude to the far away, but they are always about the here and now, about seizing the moment and searing it into memory.

Being osmotic, fashion absorbs and assimilates the stream that surrounds it. Dior introduced voluminous fabrics into women's dresses after World War II, at the same time that car manufacturers were expressing similar exuberance in lavish chrome adornments. Mary Quant miniskirts would be hard to understand without the swinging sixties and the Pill.

Clothes tell us about so many things that they have been likened to a language with vocabulary (dress, jacket, trousers), modifiers (trims, belts, scarves), slang (the baseball cap worn backward), dialects (the endless permutations of street styles), foreign accents (perfume from France, hat from England), expletives (the fuck you! of huge boots), and personal flourishes (no one can explain Cher). As with language, the expletives and slang eventually become melded into the mainstream—Doc Martens and backward caps do not make any statement these days if worn by a young person. It is easiest to understand the sartorial language of our own social group, but sometimes difficult to decode dialects and foreign languages. In every group there are the fashion originals, whose dress is witty, elegant, and even poetic, the conformists, who dress in dull but serviceable prose, and the inarticulate. As writer Alison Lurie remarked, "Even when we say nothing our clothes are talking noisily to everyone who sees us. . . . To wear what everyone else is wearing is no solution to the problem, any more than it would be to say what everyone else is saying. . . . We can lie in the language of dress, or try to tell the truth; but unless we are naked and bald it is impossible to be silent."

Sex

Fashion may chatter about many things, but the conversation is mainly about sex and status. That fashion is about sex is obvious, and even the designers of the fashion vanguard agree. "Men and women both, to an extent, get dressed to get laid," said British designer Katherine Hamnett. "Fashion is all about mating. . . . Think about an eighteen-year-old. And that energy trying on twenty different T-shirts before going out—to them it is so important. . . . True fashion obsession is something to do with sex," said Gucci designer Tom Ford.

We use fashion to make us look younger, taller, and richer, to appear unblemished and unwearied. It gives us an arsenal with which to attack a bad hair day. In other words, fashion increases our mate value. Manipulation of honest signals works even in the animal world—as we saw, birds given extra tail feathers and brighter plumage were more successful at mating than their naturally attired competitors. It works well for humans too. Fashion is a giant business, in part devoted to false advertisement. Its claims of success in thwarting honest signals are overhyped, but they work to some extent. And in the hands of a gifted artist, such as makeup artists Way Bandy or Kevyn Aucoin, adornment can be truly transformative.

Some anthropologists believe that the original purpose of clothing was to call attention to the erotic zones of the body, not to hide them. The latter function arose later in the West, with sexual Puritanism. Primitive art and body decoration highlight the sources of fertility, and as late as the early sixteenth century Christian art emphasized the genitals of Christ. After that, loincloths and fig leaves appeared, and clothing became the guardian of the body, the inhibitor of unchaste thoughts.

But clothes tend to add sex appeal, even when their goal is to diminish it. In his satirical novel *Penguin Island*, Anatole France describes what happens when a missionary decides to clothe his newly converted female penguins. The male population is aroused.

As the monk Magis notes: "You must admire how each of them advances with his nose pointed toward the center of gravity of that young lady now that this center is veiled in pink. . . . I myself reel at this moment, irresistibly attracted toward that penguin."

Clothing takes our natural attractors and makes their message stronger. They make necks appear longer, breasts larger, shoulders wider, waists trimmer, hips curvier, feet smaller, and legs leggier. Clothes incite curiosity about the covered parts, and invite our imagination to fill in the gaps. They also reveal the body, selectively. J. C. Flugel, in his book on the psychology of clothes, suggested that styles offer a "shifting erogenous zone." Even the most modest attire gives some glimpse of the body, and this differentiates it from clothing meant to specifically advertise sexual unavailability (the nun's habit, the chador of an Islamic woman). The kimono is pulled back to reveal the nape of the neck. In the ever changing Western world of fashion, one generation shows its legs, another its chest, another its bare back, keeping sexual interest alive and focused, making the body appear to have endless erotic possibilities.

Revisionist history would suggest that even the crinoline was a nineteenth-century tool of seduction. Women considered the steel-hooped cage liberating since it freed them from the pounds of petticoats they had been wearing and allowed them to move their legs easily. According to fashion historian James Laver, the crinoline as a preserver of morality was "a hollow sham. . . . We are apt to think of the structure as solid and immovable; but, of course, nothing was further from the truth. The crinoline was in a constant state of agitation, thrown from side to side. . . . It swayed now to one side, now to the other, tipped up a little, swung forward and backward." It even shot upward. Swaying, undulating, rustling, always on the verge of revelation, the crinoline, in its nineteenth-century way, was about sexual provocation.

Early twentieth-century designers like Paul Poiret and Madeleine Vionnet liberated women from corsets and gave them less constructed, more pliant clothing. In doing so, they paved the way for this century's biggest fashion news—the unveiling of the body. Even Poiret's most beautiful models were not expected to have flawless bodies. One model's breasts "had to be rolled up like pancakes in order that they might be packed into her majestic bodice." Today, the models must have majestic bodies, even if it means by surgical implant. Fashions now require a good body, which they glorify. But bring fashion an out-of-shape or overweight body, and fashion will magnify its flaws.

Status

Sex is only part of the fashion story. After all, no one would confuse *Vogue* and *Harper's Bazaar* with *Playboy*. *Vogue* began in 1892, and its first issues were filled with society ladies, such as Gertrude Vanderbilt Whitney, wearing their own clothing. The magazine magnate Condé Nast bought *Vogue* in 1909 and, with his editor Edna Woolman Chase, he made it a magazine reflecting a "permeable, rootless, and democratic elite of looks, talent, image, money and success." As Kennedy Fraser noted, he understood "the perpetual pebble of dissatisfaction even in expensive handmade shoes." *Vogue* and its rival, *Harper's Bazaar*, are all about that pebble.

The fashions they feature are as much products of social competition as the finest bird feathers or the sweetest bird song. They reflect people trying to outdo one another in a game of "Watch me!" and driving each other to excess. They show the restless, shifting rules of people always raising a bar between themselves and those clamoring to take their places at the top. This is what can turn fashion into a snobbish, exclusionary business. Only through an elaborate code of rules can those at the top defend their position. To be truly in the know, one needs dedication and discernment, as well as time and money. Lapses expose the arrivistes and posers. A fashion faux pas is not an aesthetic gaffe so much as a social and moral one. The code makes people care tremendously about tiny details of cut (size of lapels, flare of pants), color, and material, and makes them abandon an article of clothing once the style becomes passé, though it may be in mint condition and has cost a fortune.

A fashion that is in one season and out the next shows us what pure status effects look like. Stripped of social meaning, the garment looks worthless, even ridiculous. Competition may drive fashion to excess and ignite fashion crazes, but the pursuit of fashion is not frivolous or silly. The game may be frenzied but the players are operating rationally. They know that clothes are valuable currency in the social arena. They show that we are ahead of the pack or, at the very least, not left behind.

Fashion Is Born

Desmond Morris called humans naked apes. Fittingly, we are also the only clothed ones. Shown a mirror, many apes will investigate their teeth and other areas they cannot usually see. Female chimps will turn around to see their pink rumps. They are interested in their appearance, but they adorn it in a haphazard way, perhaps carrying a dead mouse around on their shoulders. But that is as far as it goes. They do not use adornments to signal their status and their social aspirations.

Humans began doing so thousands of years ago, around the same time they began creating paintings in caves. In graves north of Moscow, which historians estimate to be about twenty-eight-thousand years old, explorers have found thousands of drilled ivory beads outlining the sleeves and legs of clothing they once decorated. The body of a sixty-year-old man was found with the remains of a beaded cap, ivory bracelets, and 2,936 beads arranged in strands over the body. Beside him are a boy and a girl, each covered with thousands of smaller beads. The boy and man but not the girl wear pendants or animal teeth, suggesting that even then clothing was different for the two sexes. Eyed needles have been found in upper Paleolithic cultures of Central Europe. Made from mammoth ivory, or from the bones of birds, fish, or reindeer, and using animal tendon as thread, the eyed needle functioned as it does today, piercing with its point and drawing thread through the hole. The oldest ones found in quantity are in caves thirty to forty thousand years old.

Although people began wearing clothes tens of thousands of years ago, historians date the birth of "fashion" to fourteenth-century Europe. Until then people tended to wear what previous generations had worn with only small changes. The toga, the tunic, the sari, and the kimono are examples of garments that have survived for thousands of years. Nefertiti could have met Cleopatra and been reasonably dressed although they lived a thousand years apart.

In fourteenth-century Europe people with wealth shed their loose-fitting clothes for ones that were cut and held together with buttons and laces, and showed off their physical attributes. Men's clothing got its inspiration from military dress (as men's fashions have continued to do for centuries). The armor of the knight was fitted to the body, and under it the knight wore a tight quilted and padded garment on his torso, and form-fitting hose. The undergarments of the knight became the pourpoint or doublet (a quilted garment, padded with cotton or other material, worn as a close-fitting jacket) and hose, and the padded torso and hose remained the staples of men's clothing for centuries.

Women wore long dresses with trains that flowed behind them. The bodices of their gowns had plunging necklines and were worn tight with welted seams to pull in the waist. Wide belts called bandiers separated the tight bodice from the voluminous skirt beneath. Decorations appeared on the edges of capes and hoods and on the ends of sleeves and hems, and dozens of tiny buttons were placed on the sleeves. The clothing was dramatic and individual, and people showed an appetite for continual innovation and variation.

Historians link the emergence of fashion to the new monetary power of the businessmen, bankers, merchants, and traders. Feudal society had drawn strict demarcations around wealth and status, but by the mid-fourteenth century in England, Italy, Germany, and France the feudal system was coming to an end. With the rise of trade and commerce and towns, a new class had emerged with spending power which they used to emulate the nobles. Fashion offered a visible sign of prosperity and indicated social aspiration. As cultural historian Stephen Bayley has said, "It is the prosperous but unsure middle classes who are the most voracious consumers and therefore the ones most animated by considerations of taste." Consuming products "flatters, enhances and defines people in often wonderful ways," economist Juliet Schor has written, but it also "takes over their lives."

Consumption, Waste, and Leisure

Economist Thorstein Veblen provided the seminal analysis of how people use their clothes to establish social position in *The Theory of the Leisure Class*, written in 1899. Arguing that "it is not sufficient merely to possess wealth or power," he defined the rules by which power and wealth must be displayed. His most famous coinage was *conspicuous consumption*, or the amassing of valuable things. Valuables were defined as things that were rare, difficult to find, or which reflected hours of skilled labor (by other people). When Queen Elizabeth I ascended the throne at age twenty-five in 1558, she was dressed in ermine, brocade, silk, and pearls. When the model Stephanie Seymour recently married millionaire art collector Peter Brant she wore a dress designed by Azzedine Alaia that required over nine hundred hours of work and had forty-eight thousand tiny mirrors hand-sewn on it.

One's clothes must also give evidence of *conspicuous leisure.* By leisure, Veblen meant enjoying activities that do not involve making money or producing anything useful. Clothing often reflects high-status pastimes such as hunting, golf, yachting, or polo. The riding coat and hat from the English hunter were the inspiration for the top hat and tails of evening wear, the brass buttons and blazers of the yachting world became acceptable sportswear on land, and the cardigan sweater and the polo shirt migrated from the playing field to the home. Today we have what has been dubbed "patagonia couture," clothing derived from scuba diving or snowboarding or mountain climbing.

Since the role of warrior has always been associated with the upper class, the vestments of battle are also approved status symbols. From the military have come the trench coats, epaulets, tank watches, peacoats, and khakis, among other things. As Quentin Bell has written, "The importance of war and sport to the student of dress lies in the fact that these occupations have at various times been the chief and almost the only active employment of entire castes and classes."

Another way to convey that one leads a life of leisure is to wear fabrics that require high maintenance. Linen is a good example, because it is a high-prestige fabric that wrinkles the moment you put it on. The satin brocade and embroidered shoes of the seventeenth-century French aristocracy showed that their women never walked in the mud. They did not have to: their sedans were brought directly into public rooms at Versailles. Today's diaphanous slip dresses do not look as if they could stand up to any challenge at all.

Leisure is most obviously displayed by fashions that make labor impossible. Chinese aristocrats wore long fingernails to show that they did not perform manual labor. As a recent fashion reporter noted, "High heels are for those who pay other people to do their walking for them—to the dry cleaner, to fetch a cab, to pick up lunch." In Baldassare Castiglione's 1538 *The Courtier*, he mentions *sprezzatura* and *disinvoltura*, meaning hauteur, ease, and effortlessness, essential qualities of the courtier. The idea is never to look hurried or strained.

Finally, status is shown by *conspicuous waste*, basically meaning that one is not afraid to spend because there is always more where that came from. Charlemagne owned eight hundred pairs of fine gloves, at a time when gloves were difficult to produce and to clean. The Duke and Duchess of Windsor had their toilet paper hand cut, and their footmen served their dogs from silver bowls. The early twentieth-century New York socialite Rita de Acosta owned eighty-seven black velvet coats which differed only in their lace trims. The Italian shoemaker Ferragamo once sold seventy pairs of his shoes to Greta Garbo in a single outing, and one hundred to the Maharani of Cooch Behar, who sent him pearls and diamonds to adorn them.

There is probably no better example of conspicuous consumption, waste, and leisure than the seventeenth-century court of Louis XIV at Versailles. His courtiers vied with one another for favor, and struggled endlessly at court for prestige and position. It is said that the favored few paid a huge price to receive a warrant authorizing them to wear a jerkin similar to the one worn by the king—blue moiré lined in red and embroidered with silver. The courtiers were a small group of people who cared about nothing but each other and who had nothing better to do than outdo one another. It is in such an atmosphere that competition flourishes and fashions get driven to extremes.

To set the tone of this court, Louis XIV made getting dressed in the morning a public spectacle that was called the *grand levée*. After he awoke in the morning and was handed his wig, Louis admitted about one hundred courtiers to his room. With the help of his assistants, he put on his understockings, his knee breeches, which had silk stockings attached to them, his shoes, which were ornamented with diamonds on the buckles, and his garters (these the king fastened himself). A this point he paused and had breakfast. Next, two valets held up his dressing gown so that he could put on his shirt shielded from the audience. He then put on his sword, his relics, his cravat, his lace handkerchief, his coat, hat, gloves, and cane. The order and etiquette were precise down to who presented the king with which garment and when. The levée ended once the king was dressed.

It is said that the aging Louis was barely able to walk during his last ceremonial appearance in 1715, so heavy were his garments. In a fine example of the shifting status of emblems of status, the huge blond wigs that Louis favored, and that all the courtiers wore, were later sold in bins on the street as rags and dust mops.

Outrage

Quentin Bell added a fourth rule of status display: *conspicuous outrage*. Only high-status individuals can afford the pleasure of not pleasing. Protected by social position, they are free to create their own rules, and have been among the patrons and creators of avant garde fashion and art. British aristocrat Stephen Tennant, the inspiration for Evelyn's Waugh's Sebastian Flyte in *Brideshead Revisited*, was photographed by Cecil Beaton in the twenties wearing a pin-stripe suit, a leather jacket, and lipstick. As his biographer John Hoare said, "He looked as though he had come out of the 1990's rather than the 1920's." Lady Ottoline Morrell, lover of Bertrand Russell and wife of a member of Parliament, wore clothes that looked as if they came out of "a painting by Velasquez." Although made from sumptuous materials, the dresses were often unlined, barely fastened, and merely tacked into position. They were more like "theatrical props" than clothes, designed "not for comfort but to stir the imagination of the viewer." Many of the top couturiers today have aristocratic "muses," such as Amanda Harlech, the muse for Karl Lagerfeld and former muse of John Galliano, and Isabella Blow, the former muse of Alexander McQueen of Givenchy. Harlech is Lady Harlech, and Blow, who came to the 1998 Paris couture shows wearing a dress that "looked like an open parachute," is the granddaughter of Sir Delves Broughton, the man acquitted of the murder of Lord Erroll in Kenya in 1941 (made famous by the book and movie *White Mischief*).

To fear being in "bad taste" or looking ugly or vulgar are middle-class concerns. The middle class are fashion followers, the most conservative of whom are dragged into wearing a style only because it has become so prevalent that it would be nonconformist *not* to. The upper classes only fear being mistaken for their middle-class imitators, which is why they abandon a fashion as soon as it is adopted by them. As fashion editor Diana Vreeland once advised a junior editor, "Never fear being vulgar, just boring, middle class or dull." Fashion begins in outrage, ends in mainstream acceptance, and reemerges only later when the imitators have long gone.

Outrageous clothes belong only on those with the right attitude. Compare the middle class beauty queens, the Miss Americas who smile as they parade in their evening gowns and bathing suits, who talk about social issues, travel with chaperones, and exude sincerity and earnestness, with the high-fashion model who smokes and parties, looks like a heroin addict, won't get out of bed for less than ten thousand dollars a day, and rarely smiles. Models have the world on a string, and they show it. Their job is to represent the elite, to astonish, to provoke envy, but not to please.

Crowd Control

For the first five hundred years of fashion's existence, sumptuary laws (that regulated or limited expenditures) were issued to limit everything from who could wear which fabric to how wide a skirt or how long a shoe could be. The upper classes attempted to reserve the exclusive right to wear longer shoes, bigger ruffs, wider skirts, higher heels, and scandalously short doublets. But the effort was a colossal failure. The middle class was determined to walk, literally, in the shoes of the upper class.

Sumptuary laws merely accelerated the pace of fashion. When the middle classes were forbidden to wear one style or color, they found ways to circumvent the law and in doing so invented new fashions. When Venetians were allowed to wear no more than one pearl necklace that adhered to the base of the neck, they complied by wearing a single strand that fell in layers of pearl bars down to the hem of the dress. In eighteenth-century Japan sumptuary laws restricted the use of satins with gold threads, dappled dyes, and embroidery to the samurai class. The wealthy merchant class responded by wearing dark-colored kimonos with sumptuous fabrics hidden in the lining. The minimalist aesthetic came to be known as *iki*.

More often, sumptuary laws were flouted. At the end of the fourteenth century and for much of the fifteenth century in France, Burgundy, Italy, England, and other countries, men wore long, pointy shoes called poulaines. Some have claimed that poulaines began when a nobleman needed a shoe to wear while he suffered from an ingrown toenail. Eventually all men except laborers wore them. Numerous sumptuary laws limited their length to six inches, and then to two inches for all but the nobles. But some men wore shoes so long that they had to be held by a chain to the leg so that they would not trip. Eventually the shoes became unfashionable and were replaced by square-toed shoes.

In 1476, Venetian law established a magistrate called the "Head of Pomp" whose job it was to legislate luxury. But flouting the law became a matter of pride. It was called "paying the pomp" and meant that the person was willing to pay a fine if discovered wearing the forbidden luxury. The urge to imitate the upper class even reached the clergy. The Venetian Episcopal synod of 1438 reminded the clergy to refrain from wearing the popular style of "small jackets . . . that were so short that they clearly revealed their navels" and tights "so tight that every morning they are sewn into them and unsewn." The Church ordered their wayward monks and priests to cut their hair short and round and to adopt a cassock.

Between 1706 and 1709 the Magistrate of Pomp came up with what he thought was a way to stop the emulation of the upper classes: he ordered that married noblewomen or citizens wear only black. What happened? The highest social classes obeyed the edict but only on official occasions. Much worse, women who were not noble began to dress in black. When embroidered and brightly colored fabrics were no longer status symbols, the lower classes lost interest in them.

Why did the state try to legislate such seemingly trivial behavior? As sociologist Erving Goffman has said, objects are status symbols if their purchase indicates membership in a particular status group. If outsiders buy these objects, they lose their value as status symbols. In modern times purchase is controlled through exorbitant price, elite and intimidating places of purchase, and social norms. Before the eighteenth century such exclusivity was controlled by law.

The Cult of the Designer

The state finally lost interest in fashion. The last sumptuary laws in England were issued in 1643 by Charles I and repealed at his execution in 1648. In France the last ordinances against luxury came in 1720. Following the French Revolution, the 1793 Convention made it official: "No person of either sex can force any citizen, male or female to dress in a particular way. . . . Everyone is free to wear the garment or garb suitable to his or her sex that he or she pleases." It barred cross-dressing, but little else.

Fashion became democratized in the nineteenth century. With the invention of the sewing machine, ready-to-wear clothing, and the department store, the average person was given a range of clothing choices. European men distanced themselves from the sartorial excesses of the fallen nobles and aristocrats and adopted sober suits distinguished by fine tailoring and exquisite detail.

It was at this time that the designer was born, an artist who lent a new source of prestige to the highest fashion. Dresses for the wealthy had always been created by humble hirelings, women who followed their clients' dictates. Charles Frederick Worth changed all that when he opened his shop in Paris in 1858 to sell "ready-made dresses and coats, silk goods, top-notch novelties" which his wife modeled in his salon. The Empress Eugénie, wife of Napoleon III, bought one of his dresses, and by the 1880s and 1890s the house of Worth was making dresses for nobles, aristocrats, and popular entertainers like Sarah Bernhardt. Worth's made-to-measure crinoline creations were a world away from the ready-made clothing in the department stores. Those were made by artisans and machines, his were hand-sewn by an artist. By insisting on creative control, Worth elevated the dressmaker from anonymous craftsman and tradesman to art star, and soon dozens of imitators opened shop in Paris.

For the next hundred years Parisian couturiers provided the authoritative voice in fashion. They set the standard, and the rest of the world scurried to imitate. Worth established haute couture as an art form and Paris nurtured its growth. The Chambre Syndicale de la Couture Parisienne was established in 1885 and tightly regulated the art of "high sewing": everything is made by hand, created at the atelier, and fitted with exactitude. Inspired by a swatch of fabric, a glimpse of a woman on the street, an image in a history book, a painting, the ballet, travel, or by their imagination, Paul Poiret, Madeleine Vionnet, Coco Chanel, Christian Dior, Cristóbal Balenciaga, and Hubert Givenchy created some of the world's most poetic and inspired clothing.

Haute couture is an evanescent art, dependent on the grace of its wearer. As such it is similar to ballet, which also depends on sylphlike bodies to transmit its beauty. As Bruce Chatwin noted, when he visited the ninety-six-year-old Madeleine Vionnet in her home in Paris, "She sees herself as an artist on the level of, say, Pavlova. She was single-minded in the pursuit of perfection, and even her exemplary common sense is tinged with a streak of fanaticism. The workmanship of her house was unrivalled. . . . She would handle fabric as a master sculptor realises the possibilities latent in a marble block." It is said that Balenciaga could take apart a dress with his thumbs, and that he would work for thirty-six hours without sleep to get an armhole just right.

Couture belongs to the age of the transatlantic cruise and not the Concorde. Clients must come for three to five personal fittings and may wait months for the final product. Despite the fame of haute couture and the visibility of its designers in the fashion magazines, today no more than three thousand women a year throughout the world buy the made-to-measure clothes produced in its twenty-one houses. When Dominick Dunne reported on the Paris couture shows in 1998, he was told that the buyers were mainly "women you've never heard of whose husbands have made fifty million or so in the last few years." The clothes are fabulously expensive. T-shirts can cost six thousand dollars, suits thirty thousand, and evening gowns a quarter of a million. But for the women who wear them "the couture is a necessity. . . . It's having something no one else has."

The Designer Label

Some of the clothes shown at the Paris couture shows may never be worn. A couture house could not exist by couture alone, it is too costly. Most of the money is made in selling scents, cosmetics, handbags, and sunglasses, in jeans lines, and in ready to wear. For many of the buyers of these lower-priced items, high fashion is not about owning something that no one else has. It is about owning something that rich people have, and the more clearly the item demonstrates its connection to high-status people the better. The designer's logo is worn like heraldry, and has become an essential part of the costume's appeal. Tommy Hilfiger says, "I can't sell a shirt without a logo," whereas Donna Karan once lamented, "All of this talk about brands! What ever happened to the cashmere?"

About twenty-five percent of designer sunglasses and watches in the United States are fakes, and counterfeiting is now a two-hundred-billion-dollar business. Street vendors in New York sell knockoff Rolex and Tag Heuer watches, Louis Vuitton and Gucci and Prada handbags, and Tommy Hilfiger and Ralph Lauren clothing. Many people would rather have these products, often made with cheap materials and meant to self-destruct, than products made from finer materials which do not have the faked status marker. The clothing in the chain stores, from Banana Republic to Sears, ape the highest fashions. They are just responding to the demands of an audience increasingly well informed about the goings on at the top of the fashion world.

High fashion is no longer the spectator sport of the few. Now, fashion is presented on television and on Internet sites. The most watched show ever on E! television was a program aired prior to the Academy Awards ceremony honoring the top U.S. films. The show's hostess, comedienne Joan Rivers, accosted the nominees as they entered, not to get them to talk about the films, but to get them to talk about their clothes.

Knowing that a lot of money can be made by having their clothes seen on actresses, musicians, and athletes, the designers rush to supply them with loaned clothing. This feeds the mass appetite for the up-to-the-minute image of fame and fortune. With the designers courting the masses, how can clothing reflect elite status, which should reflect rarity? What signifies status in a world of counterfeits?

Supermodels and Designer Bodies

To figure out the answer, you have to realize that all the money in the world won't make high-fashion clothes fit most bodies. Like Cinderella's slipper, they can't be slipped onto the feet of the ugly sisters. The viewers who tune into E! television will see bare shoulders and uplifted round breasts, flat abdomens and snake hips beneath semitransparent slip dresses and stretchy body-hugging fabrics. They will find that the new fashion superstar is the body, and that it has become the site of conspicuous consumption. You may not be able to tell the rich person from the poor with their copy black dresses and imitation watches and handbags, but chances are the rich person will be a lot thinner. Chances are that the rich person will have a body resculpted through gym workouts, advice from personal trainers, liposuction, and possibly implants. The rich person's body will be a lot more expensive to maintain, and will look it.

One sign that the designer body has become a status marker is the shift away from secrecy about plastic surgery. Instead of hiding plastic surgery, women in rich communities like Palm Springs flaunt it. Women may arrive at parties with their stitches showing and boast about the cost of the surgery. In some of the wealthiest circles there is the dress by the couturier, and the face by the elite and well-known surgeon, and the body by the high-ticket personal trainer. Like dress designers, the surgeons know their handiwork, and often their clientele can recognize whose face or breasts were done by whom.

There are probably many reasons why people seem as obsessed with having designer bodies as designer clothes. As we've seen, high-status clothing is easily counterfeited and copied. High-fashion clothes are also simply less stunning than they used to be. The aesthetic is simple and streamlined. In the 1920s, Chanel introduced the little black dress and humble fabrics, such as jersey, which had only been used for men's underwear. Paul Poiret sniffed that Chanel had invented "poverty deluxe." Today the trend away from ornamentation and toward simplicity and deluxe poverty continues. Much of the most prized clothing is created from a minimalist aesthetic.

The other striking contemporary phenomenon is the rise of the supermodel. Beginning in the 1960s, Twiggy, Jean Shrimpton, and other models became known by name. Like designers, the supermodels started out as humble hirelings, "clothes hangers," as model Lisa Fonssagrives used to call herself. The "supermodel" phenomenon is usually dated to 1990, when British *Vogue* featured Linda Evangelista, Christy Turlington, Cindy Crawford, Naomi Campbell, and Tatiana Patitz on its cover and identified them as the top models of the world. Today the best models lend status to the collections; their presence on the runway signifies that the designer has the money and clout to hire them.

The models of the 1920s, 1930s, and 1940s were not household names. But, like the models of today, they moved in wealthy and aristocratic circles. They were so thin that the photographers had to pin the clothes so that they would fit onto their slim hips. Maintaining the extreme thinness of models takes a lot of effort. It's hard to be that thin without cheating (drugs, smoking, eating disorders) and many thin people do cheat. It requires money and leisure time, and obsessive focus. It requires control over everything you eat.

Similar pressures are starting to bear down on male models, especially as they gain in status. Although model Marcus Schenkenberg is paid to advertise clothing, he is shown without it a remarkable percentage of the time. For his multi-page layout for a Calvin Klein advertisement, Schenkenberg was informed by photographer Bruce Weber, "You're not going to actually wear the pants, Marcus. . . . Maybe you'll take them off and cover yourself." Apparently the important thing to convey to potential buyers was not what the pants looked like but what Marcus's extraordinary body looked like.

Of course, the starvation diets and the long hours at the gym are worth it for male and female models. For them, the potential rewards are huge and the difference between making it and a near miss is the difference between great fortune and total obscurity. As economist Robert Frank has pointed out, modeling is a winner-take-all market—there are a small number people at the top competing for the big prizes. There are only about a dozen supermodels in the world. In such a market, tiny advantages can mean all the difference, and the difference is meaningful, earning millions versus earning almost nothing. Modeling is a high-paying profession only for the top few.

Just as it does not benefit the sports fan to have seven-foot basketball players or 350-pound football players, it does not benefit the fan of fashion to have models be ultra thin. Football would be as exciting whether there were 200-pounders versus 200-pounders or 300-pounders versus 300-pounders; all that matters is that they be evenly enough matched so there is a real contest. But as long as the 201-pounder beats the 200-pounder, the 202-pounder beats the 201-pounder, and so on, weight will escalate. Models will get taller and thinner as long as the rewards at the top are huge, and tiny advantages may make the difference. What are the rewards for the many people who emulate the models? The status of looking like a supermodel.

I doubt that we will see a reverse trend toward a preference for overweight men or women, but extreme thinness is bound to go the way of three-foot-high hair and the eight-foot-wide skirt. There is nowhere to go with it: models can't get any thinner, and fashion never stays in one place. Models will probably continue to get taller (because they can). American fashion may rediscover the buttocks as the new region to hone to just the right curvature after a long fetish with the breast. We may already be seeing a backlash against the breast implant. In the United Kingdom the *Sun* has banned from its daily page three feature photographs of models who have surgically enhanced breasts, and in the United States the newest porn magazine is *Perfect 10*. It is the world's first silicone-free pornography, whose models are guaranteed by its publishers to boast only the real thing.

Plus ça Change

Where is fashion heading? Today, Parisian designers compete with designers in Japan, London, Antwerp, and elsewhere and borrow from the creative outpourings of street fashion. People strive to defy categorization and emphasize their uniqueness, and want clothes that express this desire. Sex is considered a cultural construction, and clothing is said to be veering toward a unisex ideal.

The desire to defy categorization is taken to extremes in clubs and on the street among the younger wearers of fashion. Fashion historian Ted Polhemus calls the Tokyo clubs the "style surfing capital" of the world, a place where on any given night an "encyclopaedia of the entire history of western streetstyle" can be seen. Fashions are mixed and matched, modified, copied, subverted, bent to new meanings. Instead of a legible logo, there is a profusion of messages. One needs a lot of background to figure out what message is being conveyed by a young woman Polhemus sees on the streets of London whose clothes include "signifying references to (among other things): in-your-face, Tank girl feminism, Indian ethnicity, Swinging London, 60s futurism, sexual liberation, women of ill-repute, 70s glam, cowboys, rebels without a cause, on the road bohemians, Mods, Skinheads, Punks, Hippies, and students, as well as high-fashion elitism signalling wealth." Of course, the last casts an interesting light on all the rest.

Are we seeing the end of fashion? Will there be no more status symbols, no more overt sexuality, are we too cool to emulate anyone else? Well, probably not. In a recent issue of *The Face*, the British style magazine, Peter Lyle and Laura Craik ponder why thousands of kids are wearing silver Air max metallics and are so susceptible to any product advertised with fluorescent graphics and a drug reference. They conclude, "However much our generation might rail against the idea of a uniform, we still like to invent our own. There's something comforting about conforming, and something way too vulnerable/slaggable about deviating too far from the norm. . . . The Nineties as the cult of the individual? Nah."

After all the talk about gender bending, women are still parading down runways in slip dresses and men are still wearing long pants. We've been marking the differences between the sexes since the Paleolithic. That being said, how we mark them is an open question. No clothing item is intrinsically masculine or feminine. Men have worn silk hose, velvet shoes, and hair flowing to their shoulders. Women have easily borrowed from men's fashions. Men and women may more freely exchange and swap fashions but they'll figure out a way to play with the difference, it is part of the sizzle of sex.

Smart Clothes

What will we wear in the future? Probably not so many logos. The elite fashion world is showing signs of leaving them behind. The newest fashions from Hermès have only a tiny H stitched into the center of the buttons. Dolce and Gabbana are chucking their D&G label because it's been ripped off one too many times. As soon as the elite lose interest, we know a fashion trend is on the way out. Is haute couture dead? I doubt it. Paris won't let itself grow old. Some people were shocked when the staid houses of Dior and Givenchy hired avant-garde English designers John Galliano and Alexander McQueen to head their houses. But this is how it always has been. The elite aligns with the avant garde to create the new, and both embrace outrage.

A new word has cropped up in discussions among many of the avant-garde clothing designers—intelligent. The Japanese designers Yohji Yamamoto and Rei Kawakubo, the Belgian designer Martin Margiela, Helmut Lang, and Ann Demeulemeester have all been making clothes that make you think about the way they are constructed, their relation to the body, and our aesthetic presumptions. Margiela is the most avant-garde. His clothing is a half-made assembly of visible seams and recycled fabric. One gets to play an intellectual game with the clothes and look fabulous.

Designer Betsey Johnson predicted, "I think designers, if they're called designers, will have to be scientists in the future." Alex Pentland and his group at the MIT Media Lab couldn't agree more. They're constructing "wearables," computers so lightweight and portable that they can be incorporated into clothing and worn much as we wear watches or eyeglasses or jackets. Hidden in jewelry, hats, shoes, and even in the fabric itself are sensors and electrical connections. So far, they've constructed a musical jean jacket with a cloth keyboard embroidered with letters and numbers in metallic thread on the shoulder and speakers in the pockets. Steve Mann, one member of the group, replaced his apartment thermostat with a radio receiver that picked up signals from sensors in his underwear that automatically turned the heater on when it was cold. He calls it smart underwear. In the works are glasses that, with the help of a computer database of personal information, will recognize faces and whisper people's names to us when we've forgotten them, or will help us navigate when we are lost.

While Pentland and his group are forging ahead on the technical end, design students from Tokyo, Paris, Milan, and New York have been transforming couture by molding the bytes into beauty. At one fashion show at MIT an organza ball gown lit up with the dancer's movement, wowing the audience. Wearables are not quite wearable yet, there are design bugs to work out, aesthetic considerations to ponder, and the right application to decide on. But Pentland's idea of clothing as "a personal assistant that travels with you, anticipating your needs" has definite cachet (who can afford a valet?) as well as immense practical value. It's an appealing idea—clothes that make you smarter, and the perfect outfit for the information age.

CHAPTER EIGHT

Conclusion

Unless all ages and races of men have been deluded by the same mass hypnotist (who?), there seems to be such a thing as beauty, a grace wholly gratuitous.

—ANNIE DILLARD

But, bless us, things may be lovable that are not altogether handsome, I hope?

—GEORGE ELIOT

Human feeling is like the mighty rivers that bless the earth: it does not wait for beauty—it flows with resistless force and brings beauty with it.

—GEORGE ELIOT

Ancient ideas about beauty were stated in absolutes. "What is beautiful is good and what is good will soon be beautiful," said Sappho. "Beauty is truth, and truth beauty," said Keats. In the postmodern world, all is relative, even beauty. Beauty is solely "in the eye of the beholder," constructed by culture or fashioned from idiosyncratic preference.

I have argued, however, that there is a core reality to beauty that exists buried within the cultural constructs and the myths. All cultures are beauty cultures, and everywhere beauty has been a powerful and subversive force, provoking emotion, riveting attention, and directing action. Every civilization reveres it, pursues it at enormous costs, and endures the tragic and comic consequences of this pursuit.

Governments have enacted laws to control sumptuous dress, the Church has railed against vanity, and doctors have expressed horror at the risks and dangers people incur in the name of beauty. None of it has made a dent. Today, there is a deep cultural dissatisfaction with the focus on beauty, but the beauty business shows no signs of abating. A pursuit so ardent, so passionate, so risk-filled, so unquenchable reflects the workings of a basic instinct. To tell people not to take pleasure in beauty is like telling them to stop enjoying food or sex or novelty or love.

Every person has needs that cannot be satisfied at one time. Certain powerful emotions direct our actions and temporarily remove our doubts. The founder of artificial intelligence, Marvin Minsky, believes that the experience of beauty is one of nature's ways of temporarily turning off the mind's storehouse of "negative evidence" (the knowledge of what not to do). He says that the sight of beauty is a signal to the mind "to stop evaluating, selecting, and criticizing." Beauty is one of the few experiences in life that allow us to say no to the mind's censures. Beauty is comforting, intoxicating, and as art critic Peter Schjeldahl has written, it "bears the imprimatur of truth." Our critical brains temporarily deactivated, we do not introspect about beauty or think about other things in its presence.

Our response to beauty is a trick of our brain, not a deep reflection of self. Our minds evolved by natural selection to solve problems crucial to our survival and reproduction. To find the sight of potentially fertile and healthy mates beautiful and the sight of helpless infants irresistibly cute is adaptive. Despite the vagaries of fashion, every culture finds the large eyes, small nose, round cheeks, and tiny limbs of the baby beautiful. All men and women find lustrous hair, clear taut skin, a woman's cinched waist, and a man's sculpted pectorals attractive. Beauty is one of the ways life perpetuates itself, and love of beauty is deeply rooted in our biology.

There is something in our love of beauty that is, as cultural critic Kennedy Fraser wrote, "heroic, hopeless, human." It is a pleasure worth celebrating, as well as one worth putting in its place. Let us remember that beauty is one, but certainly not the only, pleasure button we press, and that people seeking mates place kindness one step ahead of beauty.

Signaling with More Than Looks

Of course, visual beauty is not the only way we communicate our evolutionarily important mate signals. As Margaret Mitchell wrote at the start of *Gone With the Wind*, "Scarlett O'Hara was not beautiful, but men seldom realized it. . . ."

Human seduction involves a subtle body language of invitations and rejections. Psychologist Monica Moore has catalogued the many signals that women use to convey their interest to men, signals that predict who will be approached by whom ninety percent of the time. She says that the frequency and intensity of these gestures are better predictors of which women will be approached by men than their physical beauty. Women give darting glances, toss their heads, lick their lips, flip their hair, give coy smiles, and engage in solitary dances. They parade around a room with shoulders thrown back and hips swaying. What men do to elicit these signals has not been catalogued, although one would imagine a synchronous set of signals coming the other way.

Moore's research suggests that most of the time people give signals that they want to be approached or left alone (signals that may be unconscious), and that part of attractiveness is simply inviting another person's attention. But people sometimes approach when not invited, and sometimes issue invitations when merely flirting with possibilities. Body language cannot make a beautiful person less beautiful, just less approachable, and that can be reversed with a flick of the finger. But the same flick of the finger (or more likely the hair) may call attention to a less beautiful person, who gets the opportunity to take it from there.

Voice

In the Fellini film *8 1/2* one of the characters gathers all the women he has ever been attracted to into one room. One is an airline employee who walks back and forth in her uniform, announcing flights. He had never seen her but he had heard her voice once at an airport while waiting to switch planes, and she stayed in his memory as an object of desire. David Letterman once listed the "top ten words that sound romantic when spoken by Barry White." They included "doo-hickey" and "gingivitis." Of course the soul singer is best known for "Can't Get Enough of Your Love, Babe," but the point is that he can make anything sound good.

As Darwin noted, many animals rely on cries, grunts, and other vocalizations to express their "love, rage, and jealousy." Male insects repeat the same note rhythmically through their stridulating organs to call to or attract females, and frogs and toads croak incessantly during breeding seasons. Size and pitch are correlated, and females respond most to the lowest croaks, which tend to be issued by the largest males.

There is even one primate that has sacrificed his facial beauty (at least by human standards) so that he could enhance his vocal beauty. The male proboscis monkey that lives in the forest of Borneo has a nose that is so gigantic, he has to push it aside with his hand to get food into his mouth. Females have smaller upturned noses. Philosopher of science Helena Cronin suggests that the huge nose evolved at least in part to amplify the male's call (she compares his resonant sound to a double bass). "Ludicrous as he may look to us," she writes, "perhaps he goes to these lengths to satisfy female taste."

Although we sometimes whistle, whisper, wail, or whine to each other, most of the time we just talk. People show a great deal of agreement about which speaking voices they consider attractive, although no one has figured out yet exactly what acoustic properties attractive voices possess.

Vocal attractiveness is responsible for important judgments. When we hear attractive voices we presume that the person is more likable, competent, and dominant than a person with an unattractive voice. Although visual beauty is more persuasive, there are also cross-channel effects. A beautiful person with a squawking voice will look less attractive, and a person with a wonderful voice will seem more visually beautiful.

The adult man's voice is lower and louder than that of the average woman because his vocal cords are longer and his larynx larger. Men tend to use less of their vocal range than women do, which makes their speech more monotone, or smooth. Women tend to have softer, breathier voices and use more of a range of intonation. Men's voices are usually considered attractive if they are low, slow, and smooth. This reflects and perhaps exaggerates the typical male voice.

But voices are groomed, just as faces are, and are as subject to fashion. There is considerable overlap between the male and female pitch range, and people demonstrate flexibility in where they place their vocal range. To sound more attractive, women used to exaggerate the pitch and the breathiness of their voices. For example, in one study of Japanese females, they placed their pitch as high as four hundred Hz (the pitch range of a baby's voice). Marilyn Monroe spoke in a high, breathy whisper that combined the helpless sound of a child with the sexual knowingness of an adult: it was a conspiratorial whisper that said "Help me" and offered something in return.

Starting in the 1970s, high feminine voices were considered less attractive. They were incompatible with expressing the grown-up confidence of a sexually liberated woman, and very unconvincing in the working world. Women in the public eye lowered their voices. Margaret Thatcher took voice lessons to lower her pitch after she was told that her voice sounded shrill. Cindy Crawford, Linda Evangelista, and Paulina Porizkova have all taken lessons to lower their voice pitch to sound less girlish. Supermodels may mimic youth in their appearance, but they are at least sounding more authoritative and mature.

Both sexes may try to rid themselves of regional accents. Although far less drastic, losing accents is analogous to plastic surgery to "correct" racial or ethnic features. The favored accent in the United States now is a "clean" Midwestern sound like that of television personality Diane Sawyer. It used to be a pseudo-British accent.

Smell

For most people, "body odor" is something to wash away with soap and deodorant or to mask with perfume. But every woman who enjoys wearing her man's T-shirts or sleeping on his pillow and every baby who presses his nose against his mother's body knows differently. Baudelaire's poem "Her Hair" is an ecstatic evocation of the sight, texture, but most of all the smell of his lover's hair: "I'll plunge my head—And where the twisted locks are fringed with down/Lurk mingled odors I grow drunk upon/Of oil of coconut, of musk and tar."

Smell is a sense poorly captured in language, connected as it is with evolutionarily old regions of the brain not directly involved in language. Baudelaire makes analogies to the smells of familiar things, but they are odd combinations and not altogether pleasant (tar with coconut and musk). It turns out that smells often evoke contradictory impressions and the same smell can be pleasant or unpleasant depending on its intensity. Some compounds smell like feces at high intensity and like flowers at low intensity. Musician Brian Eno conducted his own inquiries and found that "Methyl octane carbonate for example evokes the smell of violets and motorcycles. . . . Orris butter, a complex derivative of the roots of iris, is vaguely floral in small amounts but almost obscenely fleshy (like the smell beneath a breast or between buttocks) in quantity. . . . Perfumery," he concludes, "has a lot to do with this process of courting the edges of unrecognizability, of evoking sensations that don't have names, or of mixing up sensations that don't belong together."

Of particular fascination are the chemical messengers known as pheromones. These are hormonal substances secreted by the skin that directly alter the physiology or behavior of other members of the same species. They do not always have an observable smell, and are detected through a separate olfactory system from the main smell centers. For many years, scientists believed that humans did not have this secondary system, and that we either lacked the pheromone receptor—called the vomeronasal organ or VNO—or that we had one but it was a vestigial, useless organ. However, it is now clear that humans do have a VNO near the base of the nasal septum, and that we do alter our hormones and physiology in response to pheromones.

For example, women living together often synchronize their menstrual cycles and cycle together. Recently, Kathleen Stern and Martha McClintock proved that this effect was mediated by human pheromones. The researchers found that by wiping (odorless) pheromones from women's armpits onto the upper lips of other women, they could alter the timing of their ovulation and the length of their menstrual cycles. They found evidence of two pheromones: one that was emitted on the day of ovulation that delayed ovulation in other women and lengthened the total cycle, and another that accelerated the latency to ovulation and shortened the cycle length of recipients. The study suggests that women may be unconsciously controlling one another's reproductive cycles. To exactly what end, we have yet to find out. But the research swings the door open onto a powerful new avenue of communication and sexual influence.

Look on the internet and you will find many sites selling pheromones to men that promise to draw women like flies. Pheromone research has focused on androstenes, present in both sexes but more concentrated in males, and to a far lesser extent on copulins, the volatile fatty acids in women's vaginal secretions. Many people cannot smell androstenes at all; those who can tend to find the smell unpleasant. At ovulation, women find the smell neutral. Despite the lack of conscious enthusiasm for the odor, it seems to affect women's moods and behavior regardless. They feel calmer and find people more attractive. They are drawn to the smell. For example, both men and women exposed to androstenol (alcohol derived androstene) judged photographs of women to be more attractive than they did when exposed to neutral odors. In another study, women exposed to androstenol reported feeling calmer and mellower than women who had been exposed to neutral smells. When scientists have sprayed androstenone on chairs in dentists' waiting rooms and on theater seats, women use the chairs more often, while men avoid them.

In a recent study, scientist Astrid Jutte found that men reacted to a woman's copulins in much the same fashion. Although they did not rate the smells as very pleasant, they gave higher attractiveness ratings to photographs of women and samples of women's voices after sniffing copulins than after sniffing a neutral odor, and their testosterone levels increased significantly. The less attractive the woman was, the more she gained in a man's attractiveness rating when he was exposed to her natural smell. Copulins made unattractive looks less important.

Perfumes often contain pheromones both from plants and from

male animals. Since women, not men, are attracted to many of these smells (men avoid chairs sprayed with male pheromones), women may be using perfumes less as a way to lure men than as a way to please themselves and perhaps put themselves in a calm state of mind. Although there are many dubious claims now circulating about aromatherapy, it appears that some natural plant odors may also have effects on mood and physiology. For example, peppermint may have a refreshing effect and has been shown to alter brain waves and heart rate during sleep. Some particularly interesting research in this direction has been conducted in Japan by scientist Teruhisa Komori and colleagues, who find that lemon odors reduce feelings of depression and may actually help to normalize hormone levels and immune functions in people who become clinically depressed.

Much of the research on odor has focused on women, since women have a much keener sensitivity to smell than men do, a sex difference that emerges at puberty. Women's sensitivity to smell peaks during ovulation. In intriguing research being conducted by Claus Wedekind in Switzerland, women were found to be particularly attracted to men who smelled the *least* like themselves. Women on the pill showed the opposite effect: they preferred men with body odor similar to their own. The body scents that were important turned out to be the smelly by-products of our immune systems.

Wedekind's work followed up on research showing that mice prefer to breed with other mice that have dissimilar immune system genes. These MHC genes (major histocompatibility complex) are involved in recognizing invaders and transplants, in protecting biological self from nonself. Mice sniff each other's urine and choose partners that have a different set of MHC genes, presumably to avoid inbreeding and produce offspring with stronger immune systems.

Wedekind asked forty-four men to wear T-shirts for two nights (he also gave them odorless soap and had them refrain from smoking or other smelly activities). He then had women evaluate their odors. The women preferred the scent of the men whose MHC varied the most from their own (this also smelled the least like them). The shirts of men with similar MHC reminded them of fathers or brothers. They did not find them sexy.

Mothers can recognize their babies by smell alone within six hours after birth, and within days babies can recognize their mothers' distinct smells. As adults, we can recognize our own smells well enough to reliably fish out our T-shirts from a pile of others. Wedekind's research suggests that we become attracted to the people who smell the least like our family members. When we tamper with our reproductive capability, as when we use birth control pills, we also derail this mechanism. Wedekind concludes that "no one smells good to everybody, it depends on who is sniffing whom." People sometimes wonder why some beauties leave them sexually unaroused: perhaps they are sniffing something too close to home.

Not Waiting for Beauty

Visual beauty does not reign supreme in our sensual world—we are lured by beautiful voices, gestures of invitation, and sexy smells. We are even drawn to people by secretions from their hormones and immune systems that we cannot consciously detect. Looks are not everything, even in the superficial world of attraction and glances.

But we are still left with the question of how to think about beauty, or why we should be thinking about it. After all, beauty is howlingly unfair. It is a genetic given. And physical appearance tells us little about a person's intelligence, kindness, pluck, sense of humor, or steadfastness, although we think it does. As Tom Wolfe has written, "At the very core of fashionable society exists a monstrous vulgarity: The habit of judging human beings by standards having no necessary relation to their character. To be found dwelling upon this vulgarity, absorbed in it, is like being found watching a suck 'n fuck movie." The grubbiness spreads contagiously and no one wants to touch the topic.

But our squeamishness is no reason to stay away. Knowledge is power: the more we know about human nature, the better hope we have of addressing inequalities and of changing ourselves. Scientific inquiry is different from the assignment of value, and the fact that a tendency or preference is innate does not mean that culture, nurture, and circumstance cannot radically alter its expression. Our impulses are not necessarily good, but they are resistible.

The politics of beauty needs a fresh forum, free from the attacks of the beauty bashers, as well as the unthinking reverence of beauty worshipers. As Lester Bangs wrote in 1979 about another fact of life (rock music), since it is "bound to stay in your life you would hope to see it reach some point where it might not add to the cruelty and exploitation already in the world." Beauty is not going anywhere. The idea that beauty is unimportant or a cultural construct is the real beauty myth. We have to understand beauty, or we will always be enslaved by it.

How can we put beauty in perspective? Let's keep in mind what good looks tell us. They tell us something about ancestral mate value, just as cuteness tells us about helplessness. They tell us whether a person is potentially fertile, healthy, and strong and might have genes that combine well with ours to make healthy babies (at least they did so before we manipulated them to such extraordinary degrees, lowering their information value). Zeroing in on information about fertility and health and strength was important in the Pleistocene, and this is why our brains have developed acute detectors. But we must rein in these easily excitable detectors in the modern world, where we jostle with strangers every day and see and meet thousands of people. Perhaps it's best to enjoy the temporary thrill, to enjoy being a mammal for a few moments, and then do a reality check and move on. Our brains cannot help it, but we can.

People hold the tacit assumption that the beautiful must be good. It makes them feel better about their attraction to the beautiful (it's not skin deep, it has nothing to do with sex or status) and it makes the world seem just. But it denies the ambiguity and the vagaries of human nature. As psychologist Roger Brown wrote, "Why would there be 22,000 books on, for instance, the 'enigma' of Richard Wagner? The supposed enigma is that a man who wrote some music that is sublime (Parsifal), some that is noble and romantic (Lohengrin) and some that is wise and gently humorous (Meistersinger) should have been an active anti-Semite, the seducer of a loyal friend's wife (Cosima Von Bulow) and at various times a liar, cheat, politician, egomaniac and sybarite. Why on earth not? The real enigma is that experienced people who must know that trait of character and talent have complex, shifting causes, can believe or pretend to believe, that a personality must be all of a piece morally."

But in uncoupling the beautiful from the good, we shouldn't make the equal error of pairing it with the bad. One sniffs sexual prudery among the beauty bashers and a denial of men's and women's basic sexual natures. Part of male sexuality (in particular) is undeniably pleasure in looking. This is not something inherently good or bad. As feminist Karen Lehrman has written, "Allowing beautiful women their beauty may turn out to be one of the most difficult aspects of personal liberation."

Cultivating beauty costs money, takes up time, and can drain emotional resources, and we need to figure out for ourselves how much time and effort we want to give to it. Women are heavily rewarded for their looks in a way that they are not always rewarded for their other assets, and it is only natural that they put some of their resources into its cultivation. The idea that women would achieve more if they only didn't have to waste time on beauty is nonsense. Women will achieve more when they garner equal legal and social rights and privileges, not when they give up beauty. Women need more sources of power and pleasure. All women will enjoy beauty more when they can see it as one of many equally rewarded assets.

Will we ever stop worshiping youthful looks? Let's face it, it's human to want to be sexually desirable, and human not to want to look like (or be) a sexual has-been. Women of all ages are trying to "pass" as teenagers. Women were doing this in Ovid's time, we just do it better now. But in the centuries since Ovid we have changed the sexual landscape with contraception, assisted reproduction, births by postmenopausal women, gay marriage, and voluntary childlessness. None will make the appeal of young, fertile bodies any less compelling, but they should help us broaden our criteria of what beauty looks like and at least consciously update our definition of mate value. We want to see youthful beauty and we want to see it ornamented and adorned to the hilt! But we can also educate our eyes to see beauty in forms that do not automatically push the ancient gene-replicator buttons. Probably the most radical thing a fashion designer could do today would be to rethink who should be modeling their clothes.

But we will continue to revel in the obvious spender of youthful beauty. We will only make our world a drabber place by not enjoying it, as long as we are not limited to it. Being beautiful and being prized for it is not a social evil. If we are more forgiving to beauties, well, we are more forgiving to everyone who gives us pleasure whether they sing a beautiful song, give us an interesting idea, or cook a brilliant meal. As Bertrand Russell asked, what would happen if peacocks envied each other's tails and began to feel that beauty was wicked? "A really splendid tail will become only a dim memory of the past." Rather than denigrate one source of women's power, it would seem far more useful for feminists to attempt to elevate all sources of women's power.

George Eliot, a brilliant writer who was not beautiful and who suffered for it when she was young, wrote some of the most important novels in the English language. As a girl she was called "clematis" or mental beauty, a name she viewed as "a satire." She called herself hideous and ugly, and had the misfortune to fall in love with Herbert Spencer, a man who wrote tracts about the importance of physical beauty. They remained lifelong friends but he refused to marry her because of her want of beauty. But in her mid-thirties, Eliot met the love of her life and lived with him until his death. Then, as an old woman, she married a handsome man twenty years younger than herself. Meanwhile, she wrote some of the most profound novels in the English language.

When Henry James met her she was fifty years old. He wrote to his father, "She is magnificently ugly—deliciously hideous. She has a low forehead, a dull grey eye, a vast pendulous nose, a huge mouth, full of uneven teeth, and a chin and jaw-bone qui n'en finissent pas. . . . Now in this vast ugliness resides a most powerful beauty which, in a very few minutes, steals forth and charms the mind, so that you end as I ended, in falling in love with her." She conveyed an "underlying world of reserve, knowledge, pride, and power," he said, and concluded, "She has a larger circumference than any woman I have ever seen."

As women and men seek that wider circumference in their own lives, it is wise to remember Eliot's words: "All honour and reverence to the divine beauty of form! Let us cultivate it to the utmost in men, women and children—in our gardens and in our houses. But let us love that other beauty too, which lies in no secret of proportion but in the secret of deep human sympathy." We cannot wait for beauty, we must bring it forth.

Notes

CHAPTER 1: INTRODUCTION: THE NATURE OF BEAUTY
PAGE

3 *beauty . . . does not exist:* Naomi Wolf, *The Beauty Myth: How Images of Beauty Are Used Against Women* (New York: Anchor, 1992), p. 12.

4 *Many intellectuals would have us believe:* For an interesting discussion on the fate of beauty in the contemporary world, see Dave Hickey, *The Invisible Dragon: Four Essays on Beauty* (Los Angeles: Art Issues Press, 1994).

4 *"the icing on the . . . cake":* Charles Baudelaire, *The Painter of Modern Life and Other Essays.* Trans. Jonathan Mayne (New York: Da Capo, 1964), p. 3.

5 *"If everyone were cast":* Charles Darwin, *The Descent of Man and Selection in Relation to Sex* (Princeton: Princeton University Press, 1981), pp. ii, 354.

5 *passion for ornament:* Ibid., pp. ii, 338, 342.

5 *During 1996 . . . 696, 904 Americans:* From the American Academy of Plastic and Reconstructive Surgeons, *1996 Plastic Surgery Statistics.* Department of Communications, Arlington Heights, IL.

5 *400 breast implants a day:* See Doug Podolsky, "Breast implants: What price vanity?" *American Health,* March 1991, pp. 70–75; Jean Seligman, "Another tempest in a C cup: Angry plastic surgeons fight back over implants," *Newsweek,* March 23, 1992, p. 67.

6 *these drastic procedures:* Kathy Davis, *Reshaping the Female Body: The Dilemma of Cosmetic Surgery* (New York; Routledge, 1995), pp. 70, 72.

6 *In Brazil there are more Avon ladies:* James Brooke, "Who braves Piranha waters? your Avon lady!" *New York Times,* July 7, 1995, p. 3. *more money . . . on beauty . . . Tons of makeup:* Judith Rodin, *Body Traps: How to Overcome Your Body Obsessions and Liberate the Real You* (London: Vermilion, 1992), p. 13.

6   *During famines:* Robert Brain, *The Decorated Body* (London: Hutchinson, 1979), p. 186.

6   *When Eleanor Roosevelt:* Gloria Steinem, *Revolution from Within, A Book of Self Esteem* (Boston: Little Brown, 1992), p. 216.

6   *"I was frequently subject":* Leo Tolstoy, *Childhood, Boyhood, Youth.* Trans. Michael Scammel (New York: McGraw-Hill, 1964), p. 72. *Nothing has:* Ibid., p. 145.

7   *"Had our":* George Santayana, *The Sense of Beauty, Being the Outline of Aesthetic Theory* (New York: Dover, 1955), p. 4; *"that we are":* Ibid., p. 31.

7   *150 msec.:* A. G. Goldstein and J. Papageorge, "Judgments of facial attractiveness in the absence of eye movements," *Bulletin of the Psychonomic Society, 15,* 1980, 269–70.

8   *"It's when someone walks in the door":* My assistant Lauren Cooper conducted interviews with representatives from Elite, Zoli, and Wilhelmina. I thank them all for their help. This quote is from Heidi Belmann at Zoli Agency.

9   *Brigitte Bardot:* Tony Crawley, *The Films of Brigitte Bardot* (New York: Citadel Press, 1975), p. 52.

9   *most lyrical description:* James Joyce, *A Portrait of the Artist as a Young Man* (New York: Viking Press, 1971), pp. 171, 172.

10 *a two-line poem:* Ezra Pound, *Gaudier-Brzeska, A Memoir* (New York: New Directions, 1970), pp. 86–89.

10 *An ideal of beauty:* Robert N. Linscott, ed., *Selected Poems and Letters of Emily Dickinson.* (New York: Anchor, 1959), p. 168.

11 *Dürer:* Quoted in Joseph Leo Koerner, *The Moment of Self-Portraiture in German Renaissance Art* (Chicago: University of Chicago Press, 1993), p. 156.

11 *each of these individuals:* I thank Don Symons for offering this insight, as well as many others.

12 *Linda Evangelista:* Quoted in Charles Gandee, "Nobody's Perfect," *Vogue,* September 1994, p. 617.

12 *Glamour hands, "five little shrimp":* Stephen Rae, "No, you don't have to be 5'10" and gorgeous to model," *Cosmopolitan,* April 1993, p. 207.

12 *the naked body:* Kenneth Clark, *The Nude: A Study in Ideal Form* (Princeton: Princeton University Press, 1972), pp. 5, 6.

13 *Elizabeth I:* From David Piper, *The English Face* (London: Thames and Hudson, 1957), pp. 68–69.

13 *Horace Walpole:* From Gloriana Roy Strong, *The Portraits of Elizabeth I* (London: Thames and Hudson, 1987).

13 *Veronica Webb:* Robin Finn, "More than a face: Veronica Webb is a model with great aspirations," *New York Times,* June 19, 1994, Section 9, p. 1.

14 *"three-body problem":* Paul Valéry, "Some Simple Reflections on the Body." In Michel Feher with Ramona Naddaff and Nadia Tazi, eds., *Frag-*

ments for a History of the Human Body, Part 2. (New York: Zone, 1989), pp. 399, 400.

14 *painters and dressmakers:* Quentin Bell, *On Human Finery* (New York: Schocken Books, 1976), p. 51.

15 *St. Augustine (Letters)* and *Cicero (Tusculan Disputations):* Umberto Eco, *Art and Beauty in the Middle Ages* (New Haven: Yale University Press, 1986), p. 28.

15 *"order and symmetry":* Aristotle *(Metaphysics,* Book xiii) and Plotinus *(Ennead I,* sixth tractate) from Albert Hofstadter and Richard Kuhns, eds., *Philosophies of Art and Beauty: Selected Readings in Aesthetics from Plato to Heidegger.* (Chicago: University of Chicago Press, 1964), pp. 96, 142.

15–16 *"most important . . . system":* George Hersey, *The Evolution of Allure: Sexual Selection from the Medici Venus to the Incredible Hulk* (Cambridge: MIT Press, 1996), p. 44. See also Erwin Panofsky, *Meaning in the Visual Arts,* (New York: Harmondsworth, 1970), pp. 90–100.

16 *Dürer's finger:* From Koerner, p. 156.

16 *set of orthodontic indices:* Edward Angle, *Treatment of Malocclusions of the Teeth.* 7th ed. (Philadelphia: S. S. White, 1907).

17 *anthropometrist:* L. G. Farkas, T. A. Hreczko, J. C. Kolar, and I. R. Munro, "Vertical and horizontal proportions of the face in young adult North American Caucasians: revisions of neo-classical canons," *Plastic and Reconstructive Surgery, 75,* 1985, 328–338. Reviewed in L. G. Farkas, I. R. Munro, and J. C. Kolar, "The validity of neoclassical facial proportion canons." In L. G. Farkas and I. R. Munro, eds., *Anthropometric Facial Proportions in Medicine.* (Springfield, Ill.: Charles C. Thomas, 1987), pp. 57–66.

18 *beauty alone:* Thomas Mann, *Death in Venice* (New York: Bantam, 1971), pp. 72–73.

19 *With the arrival of Christianity:* See Anthony Synnott, "Truth and goodness, mirrors and masks—Part I: A sociology of beauty and the face," *British Journal of Sociology, 40,* 1988, 607–636. See also Caroline Walker Bynum, "The Female Body and Religious Practices in the Later Middle Ages." In Feher et al., *Fragments for a History of the Human Body, Part 1,* 1989, pp. 157–219.

19 *Psychoanalysis assumed:* Sigmund Freud, *Civilization and Its Discontents,* Trans. James Strachey (New York: W. W. Norton, 1961), p. 30.

19 *There is a voluminous literature on plastic surgery and psychiatry:* One of the most frequently cited early studies of male patients was by W. E. Jacobson, M. T. Edgerton, E. Meyer, A. Canter, and R. Slaughter, "Psychiatric evaluation of male patients seeking cosmetic surgery," *Plastic and Reconstructive Surgery, 26,* 1960, 356–371. Literature reviewed in J. M. Goin and M. C. Goin, *Changing the Body, Psychological Effects of Plastic Surgery* (Baltimore: Williams and Wilkins, 1981).

19 *cosmetic surgery:* John E. Gedo, *Portraits of the Artists: Psychoanalysis of Creativity and Its Vicissitudes* (New York: Guilford, 1983).

20 *analogies between cosmetic surgery and "cosmetic psychopharmacology":* Peter D. Kramer, *Listening to Prozac: A Psychiatrist Explores Antidepressant Drugs and the Remaking of the Self* (New York: Penguin, 1993).

20 *One reason is:* John Tooby and Leda Cosmides, "The Psychological Foundations of Culture." In J. K. Barkow, L. Cosmides, and J. Tooby, eds., *The Adapted Mind: Evolutionary Psychology and the Generation of Culture* (New York: Oxford, 1992), pp. 19–136.

21 *"give me a dozen healthy infants":* John B. Watson, *Behaviorism* (New York: W. W. Norton, 1925), p. 82.

21 *beauty was shunned:* Gardner Lindzey, "Morphology and Behavior." In G. Lindzey, C. S. Hall, and M. Manosevitz, eds., *Theories of Personality: Primary Sources and Research.* 2nd ed. (New York: John Wiley, 1973), pp. 280–91.

22 *Charles Darwin . . . one of many:* Sir Francis Darwin, ed., Charles Darwin, *Autobiography* (New York: Henry Schuman, 1950), p. 36.

22 *the theory of evolution:* Leslie A. Zebrowitz, *Reading Faces: Window to the Soul* (New York: Westview, 1997), p. 1.

22 *"neglect of morphology":* Lindzey, 1973, p. 290.

22 *"Culture . . . not causeless":* Leda Cosmides and John Tooby, "Introduction: Evolutionary Psychology and Conceptual Integration," in Barkow, Cosmides, Tooby, 1992, p. 3.

22 *Until the 1960s:* See Steven Pinker, *The Language Instinct* (New York: Morrow, 1994).

22 *facial expressions:* Paul Ekman, "Universals and cultural differences in facial expressions of emotion." In *Nebraska Symposium on Motivation* (Lincoln: University of Nebraska Press), 1971, pp. 207–83. Paul Ekman, *Darwin and Facial Expression: A Century of Research in Review* (New York: Academic Press, 1973). Paul Ekman, "Universality of emotional expression? A personal history of the dispute." In Charles Darwin, *The Expression of the Emotions in Man and Animals: The Definitive Edition* (New York: Oxford University Press, 1998), pp. 363–93.

23 *As poet Charles Baudelaire wrote:* Baudelaire, 1964, p. 3.

23 *Camille Paglia on beauty, Egypt, and Naomi Wolf:* Camille Paglia, *Sex, Art, and American Culture* (New York: Vintage, 1992), pp. 262–65. See also Camille Paglia, *Sexual Personae: Art and Decadence from Nefertiti to Emily Dickinson* (New York: Vintage, 1991).

23 *"the time it takes":* Leda Cosmides and John Tooby, "Evolutionary Psychology: A Primer." Unpublished ms., University of California at Santa Barbara, p. 12.

25 *beauty is as potent a social force:* David Van Praag Marks, *Human Beauty: An Economic Analysis.* Ph.D. Thesis, Harvard University, Cambridge, MA, 1989.

31   *Avedon:* From Amy Fine Collins, "Avedon," *Harper's Bazaar*, March 1994, p. 287.

31   *"Beauty is not instantly . . . recognizable":* Robin Tolmach Lakoff and Raquel L. Scherr, *Face Value: The Politics of Beauty* (Boston: Routledge and Kegan Paul, 1984), p. 30.

32   *Infants show preferences:* J. H. Langlois, L. A. Roggman, R. J. Casey, J. M. Ritter, L. A. Rieser-Danner, and V. Y. Jenkins, "Infant preferences for attractive faces: Rudiments of a stereotype?" *Developmental Psychology*, *23*, 1987, 363–369; J. H. Langlois, J. M. Ritter, L. A. Roggman, and L. S. Vaughn, "Facial diversity and infant preferences for attractive faces," *Developmental Psychology*, *27*, 1991, 79–84. See converging evidence in C. A. Samuels, G. Butterworth, T. Roberts, L. Graupner, and G. Hole, "Facial aesthetics: babies prefer attractiveness to symmetry," *Perception*, *23*, 1994, 823–831.

32   *Prefer . . . symmetry:* M. H. Bornstein, K. Ferdinandsen, and C. Gross, "Perception of symmetry in infancy," *Developmental Psychology*, *17*, 1981, 82–86.

32   *prefer consonant to dissonant music:* M. R. Zentner and J. Kagan, "Perception of music by infants," *Nature*, *383*, 1996, 29.

32   *Within ten minutes:* C. C. Goren, M. Sarty, and P.Y.K. Wu, "Visual following and pattern discrimination of face-like stimuli by newborn infants," *Pediatrics*, *56*, *1975*, 544–549. See also D. Maurer and R. Young, "Newborns' following of natural and distorted arrangements of facial features," *Infant Behavior and Development*, *6*, 1983, 127–131.

32   *they can discriminate their mother's face:* I.W.R. Bushnell, F. Sai, and J.T. Mullin, "Neonatal recognition of the mother's face," *British Journal of Developmental Psychology*, *7*, 1989, 3–15.

32   *they begin mimicking facial actions:* A. N. Meltzoff and M. K. Moore, "Imitation of facial and manual gestures by human neonates," *Science*, *198*, 1977, 75–78.

32   *Each newborn orients immediately:* J. Morton and M. H. Johnson, "CONSPEC and CONLEARN: A two-process theory of infant face recognition," *Psychological Review*, *98*, 1991, 164–181.

32   *look almost as long at . . . eyes as . . . at the whole face:* M. M. Haith, T. Bergman, and M. J. Moore, "Eye contact and face scanning in early infancy," *Science*, *198*, 1977, 853–855.

33   *Automatic face recognition:* B. Moghaddam and A. Pentland, "Probabilistic visual learning for object representation," *IEEE Transactions on Pattern Analysis and Machine Intelligence*, *7*, July 1977, 696–710. H. A. Rowley, S. Baluja, and T. Kanade, "Human face detection in visual scenes," *Carnegie Mellon Computer Science Technical Report*, CMU-CS-95 158R, November 1995.

33   *they look back, and . . . smile:* P. Wolff, "Observations on the early development of smiling." In B. Foss, ed., *Determinants of Infant Behavior, Vol. 2*
(New York: Wiley, 1963).

33   *direction of gaze:* For a lucid discussion of eye gaze and mind reading, see
Simon Baron Cohen, *Mindblindness: An Essay on Autism and Theory of
Mind* (Cambridge: MIT Press, 1995).

34   *the human Mickey:* In David Parkinson, ed., *The Graham Greene Film
Reader: Reviews, Essays, Interviews and Film Stories* (New York: Applause,
1995), p. 40.

34   *Mickey aged in reverse:* S. J. Gould, "A biological homage to Mickey
Mouse," in S. J. Gould, *The Panda's Thumb: More Reflections in Natural
History* (New York: W.W. Norton, 1980), pp. 95–107.

34   *baby chimps:* Jane Van Lawick-Goodall, "The behavior of free-living
chimpanzees in the Gombe stream area," *Animal Behavior Monographs, 1,*
1968, 161–311.

35   *Anna Quindlen:* From N. Kelsh, *Naked Babies* (New York: Penguin Studio,
1996).

35   *slight differences in the way mothers act:* J. H. Langlois, J. M. Ritter, R. J.
Casey, and D. B. Sawin, "Infant attractiveness predicts maternal behaviors
and attitudes," *Developmental Psychology, 31,* 1995, 464–472.

36   *Babies considered ugly:* R. A. Maier, D. L. Holmes, F. L. Slaymaker, and
J. N. Reich, "The perceived attractiveness of preterm infants," *Infant Behavior and Development, 7,* 1984, 403–414; J. M. Ritter, R. J. Casey, J. H.
Langlois, "Adults' responses to infants varying in appearance of age and
attractiveness," *Child Development, 62,* 1991, 68–82.

36   *abused kids:* V. McCabe, "Facial proportions, perceived age, and caregiving." In T. R. Alley, ed., *Social and Applied Aspects of Perceiving Faces*
(Hillsdale: Erlbaum, 1988), pp. 89–95.

36   *coot chicks:* B. E. Lyon, J. M. Eadier, and C. D. Hamilton, "Parental choice
selects for ornamental plumage in American coot chicks," *Nature, 371,*
1993, 240–243. See also M. Pagel, "Parents prefer plumage," *Nature, 371,*
1994, 200.

37   *study of premature twins:* J. Mann, "Nurturance or negligence: Maternal
psychology and behavioral preference among preterm twins." In J. Barkow,
L. Cosmides, and J. Tooby, eds. *The Adapted Mind.* (New York: Oxford,
1992), pp. 367–90.

38   *paternal resemblance:* M. Daly and M. I. Wilson, "Whom are newborn
babies said to resemble?" *Ethology and Sociobiology, 3,* 1982, 69–78.

38   *Trobrianders:* B. Malinowski, *The Sexual Life of Savages in North-Western
Melanesia* (New York: Harcourt, Brace, and World, 1929).

39   *dark side to . . . father's desire for a baby who looks like him:* M. Daly
and M. I. Wilson, "Child maltreatment from a sociobiological perspective,"
Journal of Marriage and the Family, 43, 1980, 277–288.

39   *Adoptions are more successful when:* B. Jaffee and D. Fanshel, *How They
Fared in Adoption: A Follow-up Study* (New York: Columbia, 1970).

- 39   *Women's pupils:* E. H. Hess and J. H. Polt, "Pupil size related to interest value of visual stimuli," *Science, 132,* 1960, 349–350.

40   *show a young child a squirrel:* F. Keil, *Concepts, Kinds, and Conceptual Development* (Cambridge: MIT Press, 1989).

40   *preferences for particular landscapes:* G. H. Orians and J. H. Heerwagen, "Evolved responses to landscapes." In J. Barkow, L. Cosmides, and J. Tooby, 1992, pp. 555–79.

40   *environments offer prospect and refuge:* J. Appleton, *The Experience of Landscape* (New York: Wiley, 1975).

41   *Marsilio Ficino:* From "Commentary on Plato's Symposium," 1475, reprinted in A. Hofstadter and R. Kuhns, eds., *Philosophies of Art and Beauty: Selected Readings in Aesthetics from Plato to Heidegger* (Chicago: University of Chicago Press, 1964), p. 217. *Baldassare Castiglione:* From *The Courtier* (1561), quoted in Anthony Synnott, "Truth and goodness, mirrors and masks—Part I: a sociology of beauty and the face," *British Journal of Sociology, 40,* 1989, p. 622.

41   *Castiglione's ideas represent:* Anthony Synott, 1989, p. 623.

41   *Attempts to specify:* G. B. Della Porta, *De humana physiognoma* (Naples, 1586) quoted in Patrizia Magli, "The face and the soul." In Michel Feher with Ramona Naddaff and Nadia Tazi, eds., *Fragments for a History of the Human Body, Part 2.* (New York: Zone, 1989), pp. 103, 105.

42   *Apollo Belvedere . . . the totemic statue:* J. L. Koerner, *The Moment of Self-Portraiture in Renaissance Art* (Chicago: University of Chicago Press, 1993), p. 192.

42   *Petrus Camper:* See Magli, 1989; S. J. Gould, "Petrus Camper's Angle," in S. J. Gould, *Bully for Brontosaurus: Reflections in Natural History* (New York: W. W. Norton, 1992), pp. 229–40.

42   *Camper:* From Gould, 1992, pp. 235, 237.

43   *genetic diversity within races:* See L. L. Cavalli-Sforza and F. Cavalli-Sforza, *The Great Human Diasporas: The History of Diversity and Evolution.* Trans. S. Thorne (Reading: Addison Wesley, 1995). See also R. Lewontin, *Human Diversity* (New York: Scientific American, 1982).

44   *seventy-five college men were shown photographs:* M. R. Cunningham, "Measuring the physical in physical attractiveness: Quasi-experiments on the sociobiology of female facial beauty," *Journal of Personality and Social Psychology, 50,* 1986, 925–935.

44–45   *dime in the phone booth study:* R. Sroufe, A. Chaiken, R. Cook and V. Freeman, "The effects of physical attractiveness on honesty: A socially desirable response," *Personality and Social Psychology Bulletin, 3,* 1977, 59–62.

45   *two women stand by a car:* R. Athanasiou and P. Greene. "Physical attractiveness and helping behavior," *Proceedings of the 81st Annual Convention of the American Psychological Association, 8,* 1973, 289–290.

45   *help attractive people even if they don't like them:* H. Sigall and E. Aronson, "Liking for an evaluator as a function of her physical attrac-

tiveness and nature of the evaluations," *Journal of Experimental Social Psychology,* 5, 1969, 93–100.

45   *college applications . . . left in Detroit airports:* P. L. Benson, S. A. Karabenick, and R. M. Lerner, "Pretty pleases: the effects of physical attractiveness, race, and sex on receiving help," *Journal of Experimental Social Psychology,* 12, 1976, 409–415.

45   *less likely to ask good-looking people for help:* A. Nadler, R. Shapira, and S. Ben-Itzhak, "Good looks may help: Effects of helper's physical attractiveness and sex of helper on males' and females' help-seeking behavior," *Journal of Personality and Social Psychology,* 42, 1982, 90–99.

45   *"injustice of the given":* Jim Harrison, quoted by Johanna Schneller, "Brad Attitude," *Vanity Fair,* February 1995, p. 77.

46   *"democratic rhetoric of beauty experts":* Lois W. Banner, *American Beauty* (Chicago: University of Chicago Press, 1983), p. 206.

46   *Lauder's successful campaigns:* Estee Lauder, *Estee: A Success Story* (New York: Random House, 1985), p. 213.

46   *Tall people have bigger territories:* J. J. Hartnett, K. G. Bailey, and C. Harley, "Body height, position, and sex as determinants of personal space," *Journal of Psychology,* 87, 1974, 129–136.

46   *territory size and beauty:* M. Dabbs and N. A. Stokes, "Beauty is power: The use of space on the sidewalk," *Sociometry,* 38, 1975, 551–557.

46   *win arguments:* J. Horai, N. Naccari, and E. Fatoullan, "The effects of expertise and physical attractiveness upon opinion agreement and liking," *Sociometry,* 37, 1974, 601–606. S. Chaiken, "Communicator physical attractiveness and persuasion," *Journal of Personality and Social Psychology,* 37, 1979, 1387–1397.

46   *divulge secrets:* L. E. Brundage, V. J. Derlega, and T. F. Cash, "The effects of physical attractiveness and need for approval on self-disclosure," *Personality and Social Psychology Bulletin,* 3, 1977, 63–66; T. F. Cash and D. Soloway, "Self disclosure and correlates of physical attractiveness: An exploratory study," *Psychological Reports,* 36, 1975, 579–586.

47   *more at ease socially:* Reviewed in Alan Feingold, "Good-looking people are not what we think," *Psychological Bulletin,* 111, 1992, 304–341.

47   *think that they are in control:* R. Anderson, "Physical attractiveness and locus of control," *Journal of Social Psychology,* 105, 1978, 213–216.

47   *If the interviewee waited patiently:* D. J. Jackson and T. L. Huston, "Physical attractiveness and assertiveness," *Journal of Social Psychology,* 96, 1975, 79–84.

47   *women and men talked on the telephone for ten minutes:* M. Snyder, E. D. Tanke, and E. Berscheid, "Social perception and interpersonal behavior: On the self-fulfilling nature of social stereotypes," *Journal of Personality and Social Psychology,* 35, 1977, 656–666.

48   *We expect attractive people to be better at everything:* K. K. Dion, E. Berscheid, and E. Walster, "What is beautiful is good," *Journal of Personality and Social Psychology,* 24, 1972, 285–290; M. Webster, Jr. and

J. E. Driskell, Jr., "Beauty as status," *American Journal of Sociology, 89,* 1983, 140–165.

48   *expectations start in childhood:* M. M. Clifford and E. Walster, "Research note: The effects of physical attractiveness on teacher expectations," *Sociology of Education, 46,* 1973, 248–258.

48   *good-looking students often do get better grades:* M. M. Clifford, "Physical attractiveness and academic performance," *Child Study Journal, 4,* 1975, 201–209.

48   *presume . . . are intelligent:* L. A. Jackson, J. E. Hunter, and C. N. Hodge, "Physical attractiveness and intellectual competence: A meta-analytic review," *Social Psychology Quarterly, 58,* 1995, 108–122.

48   *better looking the (fill-in-the-blank):* R. M. Kaplan, "Is beauty talent? Sex interaction in the attractiveness halo effect," *Sex Roles, 4,* 1978, 195–204. M. J. Murphy and D. T. Hellkamp, "Attractiveness and personality warmth: Evaluations of paintings rated by college men and women," *Perceptual and Motor Skills, 43,* 1976, 1163–1166. D. Landy and H. Sigall, "Beauty is talent: Task evaluation as a function of the performer's physical attractiveness," *Journal of Personality and Social Psychology, 29,* 1974, 299–304.

49   *seven-year-olds:* K. K. Dion, "Physical attractiveness and evaluation of children's transgressions," *Journal of Personality and Social Pyschology, 24,* 1972, 207–213.

49   *get away with anything:* See review in Elaine Hatfield and Susan Sprecher, *Mirror, Mirror: The Importance of Looks in Everyday Life* (Albany: State University of New York Press, 1986). See also recent study by A. DeSantis and W. A. Kayson, "Defendants' characteristics of attractiveness, race and sex and sentencing decisions," *Psychological Reports, 81,* 1997, 679–683.

49   *good looks can backfire:* H. Sigall and N. Ostrove, "Beautiful but dangerous: Effects of offender attractiveness and nature of the crime on juridic judgment," *Journal of Personality and Social Psychology, 31,* 1975, 410–414. R. Mazzella and A. Feingold, "The effects of physical attractiveness, race, socioeconomic status, and gender of defendants and victims on judgments of mock jurors: a meta-analysis," *Journal of Applied Social Psychology, 24,* 1994, 1315–1344.

50   *in the sexual domain:* E. D. Tanke, "Dimensions of the physical attractiveness stereotype: A factor/analytic study," *Journal of Psychology, 110,* 1982, 63–74; G. W. Lucker, W. E. Beane, and R. L. Hemreich, "The strength of the halo effect in physical attractiveness research," *Journal of Psychology, 107,* 1981, 69–75. A. Feingold, 1992, 304–341.

50   *Four-year-olds and ten-year-olds:* K. K. Dion and E. Berscheid, "Physical attractiveness and peer perception among children," *Sociometry, 37,* 1974, 1–12. J. Salvia, J. B. Sheare, and B. Algozzine, "Facial attractiveness and personal-social development," *Journal of Abnormal Child Psychology, 3,* 1973, 171–178.

50   *Friendship is another story:* D. Krebs and A. A. Adinolfi, "Physical attractiveness, social relations and personality style," *Journal of Personality and Social Psychology, 31,* 1975, 245–253.

50   *"contrast effects":* D. T. Kenrick and S. E. Gutierres, "Contrast effects and judgments of physical attractiveness: when beauty becomes a social problem." *Journal of Personality and Social Psychology, 38,* 1980, 131–140. T. Kenrick, S. E. Gutierres, and L. L. Goldberg, "Influence of popular erotica on judgments of strangers and mates," *Journal of Experimental Social Psychology, 25,* 1989, 159–167.

51   *male guppies:* L. A. Dugatkin and R. C. Sargent, "Male-male association patterns and female proximity in the guppy Poecilia reticulata," *Behavioral Ecology and Sociobiology, 35,* 1994, 141.

51–52   *good-looking men and female orgasm:* R. Thornhill, S. W. Gangestad, and R. Cromer, "Human female orgasm and mate fluctuating asymmetry," *Animal Behavior, 50,* 1995, 1601–1615.

52   *men . . . read friendly gestures as signs of sexual interest:* A. Abbey, "Sex differences in attributions for friendly behavior. Do males misperceive females' friendliness?" *Journal of Personality and Social Psychology, 42,* 1982, 830–838.

CHAPTER 3. PRETTY PLEASES
PAGE

57   *traditional cultures:* S. Frayser, *Varieties of Sexual Experience: An Anthropological Perspective on Human Sexuality* (New Haven: HRAF Press, 1985).

57   *Holly Golightly:* Truman Capote, *Breakfast at Tiffany's* (New York: Vintage, 1993), p. 12.

58   *Adolescence is when we find out:* Mary Pipher, *Reviving Ophelia: Saving the Selves of Adolescent Girls* (New York: Ballantine, 1994), p. 55.

58   *Dustin Hoffman:* Interview by Bob Costas, 1995.

58   *Fabio:* "Beautiful dreamer," *The New Yorker,* May 17, 1993, pp. 38–39. George Wayne, "Single white narcissus," *Vanity Fair,* November 1992, p. 186.

58–59   *"at the beginning of a romance":* Elaine Hatfield and Susan Sprecher, *Mirror, Mirror: The Importance of Looks in Everyday Life* (Albany: State University of New York Press, 1986), p. 109.

59   *Hatfield's research:* (Hatfield was Elaine Walster in 1966.) In E. Walster, V. Aronson, D. Abrahams, and L. Rottman, "Importance of physical attractiveness in dating behavior," *Journal of Personality and Social Psychology, 4,* 1966, 508–516.

59   *one hundred gay men at a tea dance:* P. Sergios and J. Cody, "Importance of physical attractiveness and social assertiveness skills in male homosexual dating behavior and partner selection," *Journal of Homosexuality, 12,* 1985, 71–84.

59   *Preferences based on looks:* D. M. Buss, "Sex differences in human mate preferences: Evolutionary hypotheses tested in 37 cultures," *Behavioral and Brain Sciences, 12,* 1989, 1–14.

59   *prevalence of parasitic diseases:* S. W. Gangestad and D. M. Buss, "Pathogen prevalence and human mate preferences," *Ethology and Sociobiology, 14,* 1993, 89–96.

59 *Flamboyant signals of health:* W. D. Hamilton and M. Zuk, "Heritable true fitness and bright birds: A role for parasites," *Science, 218,* 1982, 384–387. M. Zuk, "Parasites and bright birds: new data and a new prediction." In *Bird-Parasite Interactions: Ecology, Evolution and Behavior.* J. E. Loye and M. Zuk (Oxford: Oxford University Press, 1982), pp. 317–327.

60 *"It's not that you never see a stranger across a crowded room":* R. T. Michael, J. H. Gagnon, E. O. Laumann, and G. Kolata, *Sex in America: A Definitive Survey* (Boston: Little, Brown, 1994), p. 69.

60 *roughly equivalent:* B. I. Murstein, "Physical attractiveness and marital choice," *Journal of Personality and Social Psychology, 22,* 1972, 8–12. A. Feingold, "Matching for attractiveness in romantic partners and same-sex friends: A meta-analysis and theoretical critique," *Psychological Bulletin, 104,* 1988, 226–235.

60 *More women than men diet:* See S. Hesse-Biber, *Am I Thin Enough Yet? The Cult of Thinness and the Commercialization of Identity* (New York: Oxford, 1996). R. A. Gordon, *Anorexia and Bulimia: Anatomy of a Social Epidemic* (Cambridge: Basil Blackwell, 1990). *1996 Plastic Surgery Statistics,* published by the American Society of Plastic and Reconstructive Surgeons, Arlington Heights, IL.

61 *in 1939 . . . in 1989:* See D. M. Buss, *The Evolution of Desire: Strategies of Human Mating* (New York: Basic, 1994).

61 *more "explicit consideration":* C. S. Ford and F. A. Beach, *Patterns of Sexual Behavior* (New York: Harper and Row, 1951).

61 *eight billion dollars:* Figure is for 1996. Eric Schlosser, "The business of pornography," *U.S. News and World Report,* February 10, 1997, pp. 42–50.

61 *Report of the U.S. Commission* (1970), quoted in D. Symons, *The Evolution of Human Sexuality* (New York: Oxford University Press, 1979), p. 171.

61 *sex differences in pornography:* Exact figures on who consumes pornography are difficult to obtain because of the private, often anonymous methods of purchase. See the statistics on the use of autoerotic materials (defined as X-rated movies or videos, clubs with nude or seminude dancers, sexually explicit books or magazines, sex toys and sex phone numbers) by men and women in the 1994 Sex in America Survey. Forty-one percent of men used any of the above, versus sixteen percent of women. Twenty-three percent of men viewed X-rated materials versus eleven percent of women; twenty-two percent of men went to a club with nude dancers versus four percent of women, and sixteen percent of men read sexually explicit books or magazines versus four percent of women. E. O. Laumann, J. T. Gagnon, R. T. Michael, and S. Michael, *The Social Organization of Sexuality: Sexual Practices in the United States* (Chicago: University of Chicago Press, 1994), pp. 134–141.

61 *Viva and Playgirl:* Symons, 1979, pp. 174–176.

61 *Men are much more likely to fantasize:* B. J. Ellis and D. Symons, "Sex differences in sexual fantasy: An evolutionary psychological approach," *Journal of Sex Research, 27,* 1990, 527–555.

62 *"boywatch" and "girlwatch":* From Symons, 1979, p. 182.

62 *personals ads:* K. Deaux and R. Hanna, "Courtship in the personals col-
    umn: The influence of gender and sexual orientation," *Sex Roles, 11,* 1984,
    363–375; M. N. Hatala and J. Prehodka, "Content analysis of gay male
    and lesbian personal advertisements," *Psychological Reports, 78,* 1996,
    371–374.

62 *homosexual mate preferences:* Don Symons, *The Evolution of Human Sex-
    uality* (New York: Oxford University Press, 1979).

63 *short-term relationships:* See D. M. Buss and D. P. Schmitt, "Sexual strate-
    gies theory: An evolutionary perspective on human mating," *Psychological
    Review, 100,* 1993, 204–232.

63 *more serious relationships:* Buss, 1989.

63 *age and attractiveness:* W. R. Jankowiak, E. M. Hill, and J. M. Donovan,
    "The effects of sex and sexual orientation on attractiveness judgments: An
    evolutionary interpretation," *Ethology and Sociobiology, 13,* 1992, 73–85.

64 *age of bride and groom:* Arlene Saluter and Terry A. Lugaila, "Marital
    Status and Living Arrangements: March 1996," U.S. Department of Com-
    merce, Economics and Statistics Administration, U.S. Bureau of the Census
    P20-496, 1998.

64 *If men marry a second time:* From Buss, 1994.

64 *Meryl Streep:* From Maureen Dowd, "Go ahead, make him cry," *New York
    Times,* March 26, 1995, Sec. 2, pp. 1, 27.

65 *Buss quoted:* From Janet Roach, "What's unusual about these pictures?"
    New York Times, May 22, 1994, p. 19.

65 *The best-looking girls in high school:* J. R. Udry and B. K. Eckland, "Bene-
    fits of being attractive: Differential payoffs for men and women," *Psycho-
    logical Reports, 54,* 1984, 47–56.

65 *Better-looking girls tend to "marry up":* G. H. Elder, Jr., "Appearance and
    education in marriage mobility," *American Sociological Review, 34,* 1969,
    519–533. P. A. Taylor and N. D. Glenn, "The utility of education and
    attractiveness for female status attainment through marriage," *American
    Sociological Review, 41,* 1976, 484–498.

66 *women with greater intelligence:* N. F. Marks, "Flying solo at midlife: Gen-
    der, marital status, and psychological well-being," *Journal of Marriage and
    the Family, 58,* 1996, 917–932.

66 *men with very attractive women:* H. Sigall and D. Landy, "Radiating
    beauty: Effects of having a physically attractive partner on person percep-
    tion," *Journal of Personality and Social Psychology, 28,* 1973, 218–224.

66 *"Women don't look for handsome men":* Milan Kundera, *The Book of
    Laughter and Forgetting* (New York: Penguin, 1981), p. 12.

66 *model salaries:* Joshua Levine, "We have shares," *Forbes,* March 27, 1995,
    pp. 75–78. John W. Wright, *The American Almanac of Jobs and Salaries*
    (New York: Avon, 1982), pp. 263–267; *top model salaries:* Interview with
    model Hoyt Richards, first male supermodel, January 1996.

67 *men and women mold each other:* Matt Ridley, "Why should males exist?"
    U.S. News and World Report, August 18, 1997, p. 54.

67 *"I don't feel less":* Camille Paglia, "The M.I.T. Lecture: Crisis in the American Universities." In Camille Paglia, *Sex, Art, and American Culture* (New York: Vintage, 1992), p. 264.

68 *women respond to envy:* Fran Lebowitz, "Fran Lebowitz on money," *Vanity Fair*, July 1997, p. 96.

69 *"If you desire glory":* Betrand Russell, *The Conquest of Happiness* (New York, Liveright, 1958), p. 88.

70 *"If the sixties":* Joyce Winer, "The floating lightbulb." In Patricia Foster, ed., *Minding the Body: Women Writers on Body and Soul.* (New York: Anchor, 1994), p. 47.

70 *minifetishists:* Robert J. Stoller, *Presentations of Gender* (New Haven: Yale University Press, 1985), p. 135.

72 *Symons and Profet:* From Donald Symons, "Beauty is in the adaptations of the beholder: The evolutionary psychology of human female sexual attractiveness." In P. R. Abramson and S. Pinkerton, eds. *Sexual Nature, Sexual Culture* (Chicago: University of Chicago Press, 1995), pp. 80–118.

72 *breast feeding and fertility:* S. Diaz, R. Aravena, H. Cordenas, M. Casado, and P. Miranda, "Contraceptive efficiency of lactational amenorrhea in urban Chilean women," *Contraception, 43*, 1991, 335–352. S. Diaz, M. Seron-Ferre, H. B. Croxatto, and J. Veldhuis, "Neuroendocrine mechanisms of lactational infertility in women," *Biological Research, 28*, 1995, 155–163.

72 *fertility:* J. Menken, J. Trussell, and U. Larsen, "Age and infertility," *Science, 233*, 1986, 1389–1394.

72 *Menopause is . . . cruel:* Susan Sontag, "The double standard of aging." In J. Williams, ed., *Psychology of Women* (New York: Academic Press, 1979), pp. 462–78.

73 *Lauren Hutton:* Quoted in Michael Gross, *Model: The Ugly Business of Beautiful Women* (New York: William Morrow, 1995), p. 222.

73 *shutdown . . . biologically strategic:* Jared Diamond, *Why Is Sex Fun? The Evolution of Human Sexuality* (New York: Basic Books, 1997).

73 *frozen eggs:* Personal communication, M. Jodeane Pringle.

74 *In boys' camps, rank:* R. C. Savin-Williams, "Dominance hierarchies in groups of early adolescents," *Child Development, 50*, 1979, 923–935.

75 *West Point cadets:* A. Mazur, J. Mazur, and C. Keating, "Military rank attainment of a West Point class: Effects of cadets' physical features," *American Journal of Sociology, 90*, 1984, 125–150. U. Mueller and A. Mazur, "Facial dominance of West Point cadets as a predictor of later military rank," *Social Forces, 74*, 1996, 823–850.

75 *"girls praise":* Ovid, *The Erotic Poems.* Trans. Peter Green (New York: Penguin, 1982), p. 199.

76 *men's earnings and marriage:* W. C. Wolf and M. MacDonald, "The earnings of men and remarriage," *Demography, 16*, 1979, 389–399; J. Hasky, "Social class differentials in remarriage after divorce: results form a forward linkage study," *Population Trends, 47*, 1987, 34–42; W. C. Wolf and M. MacDonald, "The earnings of males and marital disruption," Center for

Demography and Ecology Working Paper 78-14, 1978. Madison: Center for Demography and Ecology, University of Wisconsin.

76 *separation and divorce in traditional cultures:* Frayser, 1985.

76 *A woman makes her evaluations . . . more slowly:* W. G. Graziano, L. A. Jensen-Campbell, L. J. Shebilske, and S. R. Lundgren, "Social influence, sex differences, and judgments of beauty: Putting the interpersonal back in interpersonal attraction," *Journal of Personality and Social Psychology, 65,* 1993, 522–531.

77 *Model Hoyt Richards:* Personal interview, January 1966.

77 *Dennis Rodman:* From David Remnick, "Raging Bull," *The New Yorker,* June 10, 1996, p. 87.

78 *visually preferred age range for male beauty:* Symons, 1979.

78 *Dürer's self-portrait:* From J. L. Koerner, "The Moment of Self-Portraiture in German Renaissance Art (Chicago: University of Chicago Press, 1993).

78 *scorpionflies:* R. Thornhill and K. P. Sauer, "Genetic sire effects on the fighting ability of sons and daughters and mating success of sons in the scorpionfly (Panorpa vulgaris)," *Animal Behavior, 43,* 1992, 255–264.

79 *studies on income and attractiveness:* J. M. Townsend and G. D. Levy, "Effect of potential partners' physical attractiveness and socioeconomic status on sexuality and partner selection," *Archives of Sexual Behavior, 19,* 1990, 149–164. J. M. Townsend and G. D. Levy, "Effect of potential partners' costume and physical attractiveness on sexuality and partner selection," *Journal of Psychology, 124,* 1990, 371–389.

80 *"Class is the deep dark secret":* Pogrebin quoted in Sam Roberts, "When a woman earns like a man," *New York Times,* November 6, 1994, Sec. 4, p. 6.

80 *"Women represent":* Humphrey Institute, University of Minnesota, *Looking to the Future: Equal Partnership Between Women and Men in the 21st Century,* quoted in Debbie Taylor, *Women: A World Report* (Oxford: Oxford University Press, 1985), p. 82.

80 *women medical students:* J. M. Townsend, "Mate selection criteria: A pilot study," *Ethology and Sociobiology, 10,* 1989, 241–253.

80 *In surveys of college students:* M. W. Wiederman and E. R. Allgeier, "Gender differences in mate selection criteria: Sociobiological or socioeconomic explanation?" *Ethology and Sociobiology, 13,* 1992, 115–124.

81 *"laborious learning":* Immanuel Kant, *Observations on the Feeling of the Beautiful and the Sublime.* Trans. J. T. Goldthwait (Berkeley: University of California Press, 1960), p. 78

81 *divorce rates soar:* Helen Fisher, *Anatomy of Love: The Mysteries of Mating, Marriage, and Why We Stray* (New York: Fawcett Columbine, 1992).

81 *gain from trades:* G. S. Becker, *A Treatise on the Family* (Cambridge: Harvard University Press, 1981).

81 *"alters the nature of the marital bargain":* M. M. Sweeney, "Remarriage of men and women: The role of socioeconomic prospects," *Journal of Family Issues, 18,* 1997, 479–502. M. M. Sweeney, "Women, men and changing

families: The shifting economic foundation of marriage," *Center for Demography and Ecology, Working paper No. 97-14.* Madison: University of Wisconsin, 1997.

81 *new "trophy" wife:* Ralph Gardner, Jr., "Married to the Market," *New York,* June 15, 1998, pp. 24–29, 58.

82 *Marjorie Garber:* Quoted in David Berreby, "Your mom wears combat boots," *New York Times,* March 9, 1997, p. 4.

82 *Men are . . . spending 9.5 billion dollars:* Alan Farnham, "You're so vain," *Fortune,* September 9, 1996, 66–82.

83 *beauty and income:* D. S. Hamermesh and J. E. Biddle, "Beauty and the labor market, *American Economic Review, 84,* 1994, 1174–1194; T. F. Cash and R. N. Kilcullen, "The aye of the beholder: Susceptibility to sexism and beautyism in the evaluation of managerial applicants," *Journal of Applied Social Psychology, 15,* 1985, 591–605; I. H. Frieze, J. E. Olson, and D. C. Good, "Perceived and actual discrimination in the salaries of male and female managers," *Journal of Applied Social Psychology, 20,* 1990, 46–67, M. E. Heilman and L. R. Saruwatari, "When beauty is beastly: The effects of appearance and sex on evaluations of job applicants for managerial and nonmanagerial jobs," *Organizational Behavior and Human Performance, 23,* 1979, 360–372, p. 372.

83 *good-looking . . . are "sex typed":* L. A. Jackson and T. F. Cash, "Components of gender stereotypes and their implications for inferences on stereotypic and nonstereotypic dimensions," *Personality and Social Psychology Bulletin, 11,* 1985, 326–344.

84 *study out of Columbia Business School:* Heilman and Saruwatari, p. 371.

84 *Dress for Success:* J. T. Molloy, *Dress for Success* (New York: Warner, 1975). J. T. Molloy, *The Woman's Dress for Success Book* (New York: Warner, 1978).

84 *coworker wanted to have sex with them:* B. A. Gutek, *Sex and the Workplace: The Impact of Sexual Behavior and Harrassment on Women, Men, and the Organization* (San Francisco: Jossey-Bass, 1985).

85 *Happiness:* D. G. Myers and E. Diener, "Who is happy?" *Psychological Science, 6,* 1995, 10–19.

85 *beauty and subjective well-being:* E. Diener, B. Wolsic, and F. Fujita, "Physical attractiveness and subjective well-being," *Journal of Personality and Social Psychology, 69,* 1995, 120–129.

86 *"No instinct tells us that we have accumulated enough":* T. Miller, *How to Want What You Have: Discovering the Magic and Grandeur of Ordinary Existence* (New York: Avon, 1995), 75–76.

86 *beautiful female patients:* E. Jacobson, *The Self and the Object World* (New York: International Universities Press, 1964).

86 *"He forgets that":* Bertrand Russell, *The Conquest of Happiness* (New York: Liveright, 1958), p. 29.

86 *genes and happiness:* D. Lykken and A. Tellegen, "Happiness is a stochastic phenomenon," *Psychological Science, 7,* 1996, 186–189.

87   *"happier people tend to perceive themselves"*: Diener, Wolsic, and Fujita, p. 120.

87   *less faithful, seek a divorce:* M. Dermer and D. L. Thiel, "When beauty may fail," *Journal of Personality and Social Psychology, 31,* 1975, 1168–1176.

87   *"propitious physiognomies"*: Michel de Montaigne, "On physiognomy." In Michel de Montaigne, *Essays.* Trans. J. M. Cohen (New York: Penguin, 1958), p. 338.

88   *beauty and integrity:* A. H. Eagly, R. D. Ashmore, M. G. Makhijani, and L. C. Longo, "What is beautiful is good, but . . . : A meta-analytic review of research on the physical attractiveness stereotype," *Psychological Bulletin, 110,* 1991, 109–128.

CHAPTER 4. COVER ME
PAGE

91   *seeing "derived from touching"*: Sigmund Freud, *Three Essays on the Theory of Sexuality.* Trans. James Strachey (New York: Basic Books, 1962), p. 22.

91   *gloves . . . like "steamrollered silk arms"*: Kennedy Fraser, *Scenes from the Fashionable World* (New York: Knopf, 1987), p. 73.

91   *Flawless skin:* Desmond Morris, *The Human Zoo* (New York: Dell, 1969).

91   *Skin may be:* W. Montagna, *The Structure and Function of Skin* (New York: Academic Press, 1962). A. Montague, *Touching: The Human Significance of the Skin* (New York: Columbia University Press, 1971). M. Lappe, *The Body's Edge: Our Cultural Obsession with Skin* (New York: Henry Holt, 1996).

91   *nothing quite like skin gone bad:* William Ian Miller, *The Anatomy of Disgust* (Cambridge: Harvard University Press, 1997), pp. 52–53.

91–92   *Ruskin never consummated:* I thank Stan Sclaroff for relating the Ruskin anecdote to me. See also P. Rose, *Parallel Lives: Five Victorian Marriages* (New York: Vintage, 1984), p. 56.

92   *"coprophilic pleasure"*: S. Freud, 1962, p. 21, note 1.

92   *"The naked ape"*: Desmond Morris, *The Naked Ape: A Zoologist's Study of the Human Animal* (New York: McGraw Hill, 1967).

93   *denuding took place:* Marvin Harris, *Our Kind: Who We Are, Where We Came From, Where We Are Going* (New York: HarperPerennial, 1989).

94   *Alek Wek:* See *Elle,* November 1997, and A. Samuels, "Black beauty's new face: African model has impact on the runway," *Newsweek,* November 24, 1997, p. 68.

94   *grooming and primates:* Franz de Waal, *Chimpanzee Politics: Power and Sex among Apes* (London: Cape, 1982); *Azalea:* Franz de Waal, *Good Natured: The Origins of Right and Wrong in Humans and Other Animals* (Cambridge: Harvard University Press, 1996).

94   *maternal grooming:* S. R. Butler, M. R. Suskind, and S. M. Schanberg, "Maternal behavior as a regulator of polyamine biosynthesis in brain and heart of the rat pup," *Science, 199,* 1978, 445–447. S. M. Schanberg and

T. M. Field, "Sensory deprivation stress and supplemental stimulation in the rat pup and preterm human neonate," *Child Development, 58*, 1987, 1431–1447.

95   *maternal grooming and stress:* D. Liu, J. Diorio, B. Tannenbaum, C. Caldji, D. Francis, A. Freedman, S. Sharma, D. Pearson, P. M. Plotsky, and J. M. Meaney, "Maternal care, hippocampal glucocorticoid receptors, and hypothalamic-pituitary-adrenal response to stress," *Science, 277*, 1997, 1659–1662. R. M. Sapolsky, "The importance of a well-groomed child," *Science, 277*, 1997, 1620–1621.

95   *massage and premature infants:* F. A. Scafidi, T. Field, and S. M. Schanberg, "Factors that predict which preterm infants benefit most from massage therapy," *Journal of Development and Behavioral Pediatrics, 14*, 1993, 176–180.

95   *"most intimate time of the day":* Mary Catherine Bateson, *With a Daughter's Eye: A Memoir of Margaret Mead and Gregory Bateson* (New York: HarperPerennial, 1994), p. 51.

95   *"totally hair" Barbie:* In M. G. Lord, *Forever Barbie: The Unauthorized Biography of a Real Doll* (New York: William Morrow, 1994).

95   *money on personal care:* P. K. Francese, "Big spenders," *American Demographics*, August 1977, pp. 51–57. *global market: Asiaweek*, August 16, 1991, p. 15. *88% wear makeup:* S. Dortch, "Women at the cosmetics counter," *American Demographics*, March 1977, pp. 4, 6–8.

96   *red ochre . . . forty thousand years old:* Steven Mithen, *The Prehistory of the Mind: The Cognitive Origins of Art, Religion, and Science* (London: Thames and Hudson, 1996) C. Knight, C. Powers, and I. Watts, "The human symbolic revolution: A Darwinian Account," *Cambridge Archaeological Journal, 5*, 1995, 75–114.

96   *Egyptian cosmetics:* Richard Corson, *Fashions in Makeup: From Ancient to Modern Times* (London: Peter Owen, 1972).

97   *"Let no rude goat":* Ovid, *The Art of Love.* Trans. Rolfe Humphries (Bloomington: Indiana University Press, 1957), p. 159.

97   *pubic waxes:* M. Frankel, "Bikini-wax wars," *Cosmopolitan*, August 1997, p. 146.

98   *body builders:* Sam Fussell, *Muscle: Confessions of an Unlikely Bodybuilder* (New York: Poseidon, 1991).

98   *Clive James:* Quoted in K. R. Dutton, *The Perfectible Body: The Western Ideal of Physical Development* (London: Cassell, 1995), p. 339.

98   *only one Tarzan:* From ibid., 1995.

98   *"shaved muscle boy":* M. Signorile, *Life Outside: The Signorile Report on Gay Men: Sex, Drugs, Muscles and the Passage of Life* (New York: HarperCollins, 1997), p. 25.

98   *Tattoos:* Margot Mifflin, *Bodies of Subversion: A Secret History of Women and Tattoo* (New York: Juno, 1997). V. Vale and Andrea Juno, *Modern Primitives: An Investigation of Contemporary Adornment and Ritual* (New York: Juno, 1989). For an excellent summary of cross-cultural practices of

the body arts, see Robert Brain, *The Decorated Body* (London: Hutchinson, 1979).

98   *1990 survey:* Marilynn Larkin, "Tattooing in the 90's: Ancient art requires care and caution," U.S. Food and Drug Administration, *FDA Consumer*, October 1993.

99   *Tattooed aborigines:* Charles Darwin, *The Descent of Man, and Selection in Relation to Sex* (Princeton: Princeton University Press, 1965), pp. ii, 339.

99   *cicatrization, European missionaries:* See Brain, 1979.

99   *"definition of 'dressed' ":* Ann Hollander, *Seeing Through Clothes* (Berkeley: University of California Press, 1993), p. 83.

100   *male vanity market:* A. Farnham, "You're so vain," *Fortune*, September 9, 1996, pp. 66–82.

101   *Andreas de Laguna:* From Corson, 1972, p. 93.

101   *makeup in Heian period in Japan:* Sharon Romm, *The Changing Face of Beauty* (St. Louis: Mosby, 1992).

101   *Clinique:* Vanessa Friedman, "Planet Clinique," *Elle*, May 1998, pp. 218–220.

101   *makeup artist:* Kevyn Aucoin, *Making Faces* (Boston: Little Brown, 1997).

102   *Betsey Johnson quote:* James Servin, "Can Lipstick Change Your Life?" *Harper's Bazaar*, February 1994, *1,484 tubes a minute:* Judith Rodin, *Body Traps: How to Overcome your Body Obsessions and Liberate the Real You* (London: Vermilion, 1992), p. 13. Other lipstick facts from Corson, 1972.

102   *Martial:* Corson, 1972, p. 52; *Ovid;* Ibid., p. 50.

102   *St. Jerome:* Ibid., p. 68.

102   *Clement of Alexandria:* Richard Corson, *Fashions in Hair: The First Five Thousand Years* (London: Peter Owen, 1991), p. 57.

102   *English Parliament, British Journal The Spectator:* Corson, 1972, pp. 245, 196.

103   *Henry VII:* Lorne Campbell, *Renaissance Portraits* (New Haven: Yale University, 1990), p. 159.

103   *Elizabeth I:* Susan Bassnett, *Elizabeth I: A Feminist Perspective* (Oxford: Oxford University Press, 1988).

105   *skin color in men and women:* See P. L. Van den Berghe and P. Frost, "Skin color preference, sexual dimorphism and sexual selection: A case of gene culture co-evolution? *Ethnic and Racial Studies*, 9, 1986, 87–113. P. Frost, "Human skin color: A possible relationship between its sexual dimorphism and its social perception," *Perspectives in Biology and Medicine*, 32, 1988, 38–58. P. Frost, "Human skin color: The sexual differentiation of its social perception," *Mankind Quarterly*, 30, 1989, 3–16.

105   *skin color and pregnancy:* R. C. Wong and C. N. Ellis, "Physiologic skin changes in pregnancy," *Journal of the American Academy of Dermatology*, 10, 1984, 929–943.

105   *lighter skin in women:* V. S. Ramachandran, "Why do gentlemen prefer blondes?" *Medical Hypotheses, 48,* 1997, 19–20.

106   *cultural universal:* D. M. Jones, *The Evolutionary Psychology of Human Physical Attractiveness: Results from Five Populations.* Ph.D. Diss. University of Michigan, Ann Arbor, MI, 1994.

106   *ideal woman: The Kama Sutra of Vatsyayana.* Trans. Sir R. Burton and F. F. Arbuthnot (London: Diamond, 1996), p. 88.

106   *Japanese men value:* H. Wagatsuma, "The social perception of skin color in Japan," *Daedalus, 96,* 1967, 407–443.

106   *Caucasian students in Wyoming:* S. Feinman and G. W. Gill, "Sex differences in physical attractiveness preferences," *Journal of Social Psychology, 105,* 1978, 43–52.

106   *skin bleaching:* G. H. Findlay and H. A. de Beer, "Chronic hydroquinone poisoning of the skin from skin-lightening cosmetics. A South African epidemic of ochronosis of the face in dark-skinned individuals," *South African Medical Journal, 57,* 1980, 187–190.

107   *C. Darwin:* P. Ekman, ed., *The Expression of the Emotions in Man and Animals: The Definitive Edition* (New York: Oxford University Press, 1998), pp. 310, 325.

107   *blushing, and flushing:* V. S. Ramachandran, 1997, p. 20.

107   *anemia:* R. M. Nesse and G. C. Williams, *Why We Get Sick: The New Science of Darwinian Medicine* (New York: Random House, 1994).

108   *"The skin is . . . superficial:* M. Strathern, "The self in self-decoration," *Oceania, 48,* 1979, 241–257.

108   *face shapes:* S. Cavell, *The World Viewed: Reflections on the Ontology of Film* (New York: Viking, 1971), p. 70.

109   *the age machine:* Nancy Burson, *The "Age Machine" and Composite Portraiture* (Cambridge, MA: MIT List Visual Arts Center, 1990). See also N. Burson, R. Carling, and D. Kramlich, *Composites: Computer Generated Portraits* (New York: William Morrow, 1986).

109   *smoking:* D. Grady and V. L. Ernster, "Does cigarette smoking make you old and ugly?" *American Journal of Epidemiology, 135,* 1995, 839–842. See also V. L. Ernster, D. Grady, R. Miike, D. Black, J. Selby, and K. Kerlikowske, "Facial wrinkling in men and women by smoking status," *American Journal of Public Health, 85,* 1995, 78–82.

109   *tanning:* See "Sunlight, ultraviolet radiation, and the skin," *National Institutes of Health Consens. Statement Online,* 1989, May 8–10; 7(8); 1–29.

110   *nearly half . . . Seventy percent:* See "The price of beauty," *Economist,* January 11, 1992, pp. 25–26.

110–11   *Health magazine survey:* "The Great American Make-over, Health-Gallup Poll," *Health,* March–April, 1993.

111   *heard her husband:* Helen Bransford, *Welcome to Your Facelift: What to Expect Before, During, and After Cosmetic Surgery* (New York: Doubleday, 1997), pp. 45, 39.

111 *1996 plastic surgery statistics:* American Society of Plastic and Reconstructive Surgeons, Department of Communications, Arlington Heights, IL.

111 *cosmetic procedures:* American Academy of Facial Plastic and Reconstructive Surgery Annual Survey, 1993, Washington, DC.

111 *aging has a bigger impact:* F. M. Deutsch, C. M. Zalenski, and M. E. Clark, "Is there a double standard of aging?" *Journal of Applied Social Psychology, 16,* 1986, 771–785.

111 *estrogen and skin aging:* R. Maheux, F. Naud, M. Rioux, R. Grenier, A. Lemay, and M. Langevin, "A randomized, double-blind, placebo-controlled study on the effect of conjugated estrogens on skin thickness," *American Journal of Obstetrics and Gynecology, 170,* 1994, 642–649. L. B. Dunn, M. Damesyn, A. A. Moore, D. B. Reuben, and G. A. Greendale, "Does estrogen prevent skin aging? Results from the first National Health and Nutrition Examination Survey," *Archives of Dermatology, 133,* 1997, 339–342.

112 *women and acne:* B. L. Held, S. Nader, L. J. Rodriguez-Rigau, K. D. Smith, and E. Steinberger, "Acne and hyperandrogenism," *Journal of the American Academy of Dermatology, 10,* 1984, 223–226. E. Steinberger, L. Rodriguez-Rigau, K. D. Smith, and N. Held, "The menstrual cycle and plasma testosterone levels in women with acne," *Journal of the American Academy of Dermatology, 4,* 1981, 54–58.

113 *"smile, it's not that bad":* Mary Roach, "Beauty poison," *Health,* Jan–Feb 1994, p. 68.

114 *botox:* N. J. Lowe, A. Maxwell, and H. Harper, "Botulinum A exotoxin for glabellar folds: a double-blind, placebo-controlled study with an electromyographic injection technique," *Journal of the American Academy of Dermatology, 35,* 1996, 569–572. R. J. Koch, R. J. Troell, R. L. Goode, "Contemporary management of the aging brow and forehead," *Laryngoscope, 107,* 1997, 710–715.

114 *corrugator:* See Ekman, ed., 1998; Paul Ekman and Wallace V. Friesen, *Unmasking the Face* (Palo Alto: Consulting Psychologists Press, 1984). U. Dimberg and L. O. Lungquist, "Gender differences in facial reactions to facial expressions," *Biological Psychology, 30,* 1990, 151–159.

114 *brow lift study:* R. M. Freund and W. B. Nolan, "Correlation between brow lift outcomes and aesthetic ideals for eyebrow height and shape in females." *Plastic and Reconstructive Surgery, 97,* 1996, 1343–1348.

115 *Diderot:* From F. Pacteau, *The Symptom of Beauty* (Cambridge: Harvard University Press, 1994), p. 111.

115 *Paul Ekman:* Personal communication.

115 *Beauty and skin color:* S. Faison, "A Chinese edition of *Elle* draws ads and readers," *New York Times,* January 1, 1996, p. 49.

115 *Brazil:* D. J. Schemo, "Among glossy blondes, a showcase for Brazil's black faces," *New York Times,* October 18, 1996, p. A13.

115 *skin color:* See R. Lewontin, *Human Diversity* (New York: Scientific American Library, 1995). Marvin Harris, 1989.

116   *many scientists:* Jared Diamond, *The Third Chimpanzee: The Evolution and Future of the Human Animal* (New York: HarperCollins, 1992).

116   *racial differences:* L. L. Sforza and F. Sforza, *The Great Human Diasporas: The History of Diversity and Evolution* (Reading, MA: Addison-Wesley, 1996), p. 124.

116   *distinction between facts and prejudices:* Alan H. Goodman, "Bred in the bone?" *The Sciences,* March/April 1997, pp. 20–25.

117   *race relations in the West Indies:* H. Hoetink, *The Two Variants in Caribbean Race Relations: A Contribution to the Sociology of Segmented Societies* (New York: Oxford, 1967).

117   *Brazil race relations:* From D. M. Jones, 1994.

117   *Brazil Race:* from D. Schemo, 1996, p. A13

117   *An interesting discussion of waves of immigration and requests for plastic surgery is in:* Elizabeth Haiken, *Venus Envy: A History of Cosmetic Surgery* (Baltimore: Johns Hopkins University Press, 1997).

118   *Miss Americas:* From *Official Who's Who in Pageants,* 1986 ed. (Baton Rouge: International Productions and Publications).

118   *Barbie doll:* M. G. Lord, 1994, p. 75.

118   *Wilhelmina:* From M. Gross, *Model: The Ugly Business of Beautiful Women* (New York: William Morrow, 1995), p. 235.

118   *race and models:* "Race Bias Seen in Magazine Ads," *Invisible People,* New York City Department of Consumer Affairs, 1991.

118   *Women in Media group:* From Leora Tanenbaum, "Cruel beauty: Image and reality clash as real women debate the true picture of loveliness," *Boston Phoenix,* December 16, 1994, pp. 14–16.

119   *Europeans and descendants:* Tom Morganthau, "The face of the future," *Newsweek,* January 27, 1997, pp. 58–59. See also Tina Gaudoin, "Is all American beauty un-American?" *Mirabella,* September 1994, pp. 144–146.

119   *African-Americans spend . . . more on cosmetics:* Christy Fisher, "Black, hip, and primed (to shop)," *American Demographics,* September 1996, pp. 52–58.

119   *African-Americans and magazines:* Robin Pogrebin, "Success and the black magazine," *New York Times,* October 25, 1997, pp. B1, B3.

119   *"As long as societies are stratified:* Jones, 1994, p. 192.

120   *hair:* William Montagna and Richard Ellis, eds., *The Biology of Hair Growth* (New York: Academic Press, 1958); Wendy Cooper, *Hair: Sex Society Symbolism* (London: Aldus Books, 1971).

120   *hair flips:* Monica M. Moore, "Nonverbal courtship patterns in women: context and consequences," *Ethology and Sociobiology,* 6, 1985, 237–247.

120   *magazine survey: The Beauty Salon Study,* Glamour magazine and American Beauty Association, 1993.

120   *hair's provocativeness:* See Cooper, 1971, Corson, 1980. *Roman women:* See Judith Lynn Sebesta and Lynn Bonfonte, *The World of Roman Costume* (Madison: University of Wisconsin Press, 1994).

120   *bald men:* Thomas F. Cash, "Losing hair, losing points: The effects of male pattern baldness on social impression formation," *Journal of Applied Social Psychology, 20,* 1990, 154–167; see also Clifton Leaf, "The buzz on baldness," *American Health, 15,* November 1996, pp. 34–35.

121   *New Guinea Highlanders:* Andrew Strathern, "Dress, Decoration, and Art in New Guinea," in *Man as Art.* Photographs by Malcolm Kirk (San Francisco: *Chronicle,* 1993), pp. 15–40.

121   *Histories of hair styles:* See Richard Corson, 1991, and Mary Trasko, *Daring Dos: A History of Extraordinary Hair* (Paris: Flammarion, 1994).

123   *after Keats died:* "Hair of the drug that bit you," *Harvard Magazine,* September–October 1995, pp. 18–19.

124   *Merino wool:* From Montagna and Ellis, 1958.

125   *vertex baldness and risk of heart attack:* S. M. Lesko, L. Rosenberg, and S. Shapiro, "A case-control study of baldness in relation to myocardial infarction in men," *Journal of the American Medical Association, 269,* 1993, 998–1003. See also E. S. Ford, D. S. Freedman, and T. Byers, "Baldness and ischemic heart disease in a national sample of men," *American Journal of Epidemiology, 143,* 1996, 651–657.

126   *blondeness:* See Grant McCracken, *Big Hair: A Journey into the Transformation of Self* (Woodstock: Overlook Press, 1995), p. 63.

126   *blondes and fairy tales:* See Marina Warner, *From the Beast to the Blonde: On Fairy Tales and Their Tellers* (New York: Farrar, Straus, and Giroux, 1995).

127–28   *behavioral inhibition:* A. Rosenberg and J. Kagan, "Iris pigmentation and behavioral inhibition," *Developmental Psychobiology, 20,* 1987, 377–392. A. Rosenberg and J. Kagan, "Physical and physiological correlates of behavioral inhibition," *Developmental Psychobiology, 22,* 1989, 753–770.

128   *Drag queen:* RuPaul, *Lettin It All Hang Out: An Autobiography* (New York: Hyperion, 1995), p. 190.

129   *African-American women and hair:* See Kathy Russell, Midge Wilson, and Ronald Hall, *The Color Complex: The Politics of Skin Color Among African Americans* (New York: Anchor, 1993) Noliwe M. Rooks, *Hair Raising: Beauty, Culture and African American Women* (New Brunswick: Rutgers University Press, 1996).

129   *upper-middle-class blacks:* Shanette Harris cited in Debra Dickerson, "Not so Black-and-White," *Allure,* September 1997, p. 138.

CHAPTER 5. FEATURE PRESENTATION
PAGE

133   *Hortensia Borromeo:* From Lorne Campbell, *Renaissance Portraits* (New Haven: Yale University Press, 1990), p. 193.

133   *Queen Victoria's ring:* From Halla Beloff, *Camera Culture* (New York: Basil Blackwell, 1985).

133 *"related to so-and-so effect":* Melvin Konner, *The Tangled Wing: Biological Constraints on the Human Spirit* (New York: Harper Colophon, 1982), p. 322.

133 *"The difference in human features:* Francis Galton, *Inquiries into Human Faculty and Its Development* (London: MacMillan, 1883), p. 3.

134 *perceptions of facial attractiveness:* See Donald B. Giddon, "Orthodontic application of psychological and perceptual studies of facial esthetics," *Seminars in Orthodontics, 1,* 1995, 82–93. D. B. Giddon, D. L. Bernier, C. A. Evans, and J. A. Kinchen, "Comparison of two computer animated imaging programs for quantifying facial profile preference," *Perceptual and Motor Skills, 82,* 1996, 1251–1264.

134 *differences in facial morphology:* E. Westermarck, *The History of Human Marriage* (London: MacMillan, 1921), p. 8.

135 *Chinese people found:* C. Darwin, *The Descent of Man, and Selection in Relation to Sex* (Princeton: Princeton University Press, 1981), pp. 345, 352.

135 *Japan:* H. Wagatsuma, "The social perception of skin color in Japan," *Daedalus, 96,* 1967, 407–443.

135 *Michael Leahy in New Guinea:* In B. Connolly and R. Anderson, *First Contact: New Guinea Highlanders Encounter the Outside World* (New York: Viking Penguin, 1987).

35–36 *various reactions to skin color:* Darwin, 1981, pp. 346, 343, 351, 345.

136 *cross-racial identification of faces:* J. C. Brigham and P. Barkowitz, "Do they all look alike? The effect of race, sex, experience and attitudes on the ability to recognize faces," *Journal of Applied Social Psychology, 8,* 1978, 306–318. P. Barkowitz and J. C. Brigham, "Recognition of faces: Own race bias, incentive and time delay," *Journal of Applied Social Psychology, 12,* 1982, 255–268. P. Chiroro and T. Valentine, "An investigation of the contact hypothesis of the own-race bias in face recognition," *Quarterly Journal of Experimental Psychology, 48A,* 1995, 879–894.

137 *distinguishing men's faces from women's:* A. J. OToole, J. Peterson, and K. A. Deffenbacher, "An other-race effect for categorising faces by sex," *Perception, 25,* 1996, 669–676.

137 *In 1960 a London newspaper:* A. H. Iliffe, "A study of preferences in feminine beauty," *British Journal of Psychology, 51,* 1960, 267–273.

137 *A similar study done five years later in the United States:* J. R. Udry, "Structural correlates of feminine beauty preferences in Britain and the United States: A comparison," *Sociology and Social Research, 49,* 1965, 330–342.

137 *in psychologists' laboratories:* Reviewed in E. Hatfield and S. Sprecher, *Mirror, Mirror: The Importance of Looks in Everyday Life* (Albany: State University of New York Press, 1986); and L. Jackson, *Physical Appearance and Gender: Sociobiological and Sociocultural Perspectives* (Albany: State University of New York Press, 1992).

138 *agreement among seven-year-olds, twelve-year-olds:* J. F. Cross and J. Cross, "Age, sex, race and the perception of beauty," *Developmental Psychology, 5,* 1971, 433–439.

138 *women and men agree:* J. E. Meerdink, C. P. Garbin, and D. W. Leger, "Cross-gender perceptions of facial attributes and their relation to attractiveness: Do we see them differently than they see us?" *Perception and Psychophysics, 48,* 1990, 227–233.; L. A. Zebrowitz, J. M. Montepare, and H. K. Lee, "They don't all look alike: Individuated impressions of other racial groups," *Journal of Personality and Social Psychology, 65,* 1993, 85–101.

138 *research on Hiwi and Ache:* D. Jones and K. Hill, "Criteria of facial attractiveness in five populations," *Human Nature, 4,* 1993, 271–295. See also D. M. Jones, *The evolutionary psychology of human physical attractiveness: results from five populations.* Ph.D. diss. University of Michigan, Ann Arbor, MI, 1994. D. Jones, "Sexual selection, physical attractiveness, and facial neoteny," *Current Anthropology, 36,* 1995, 723–748.

138 *Cross-cultural studies:* E. Wagatsuma and C. L. Kleinke, "Ratings of facial beauty by Korean-Americans and Caucasian females," *Journal of Social Psychology, 109,* 1979, 299–300. J. N. Thakerar and S. Iwawaki, "Cross-cultural comparisons in interpersonal attraction of females toward males," *Journal of Social Psychology, 108,* 1979, 121–122. M. R. Cunningham, A. R. Roberts, A. P. Barbee, P. B. Druen, and C. H. Wu, "Their ideas of beauty are, on the whole, the same as ours: Consistency and variability in the cross-cultural perception of female attractiveness," *Journal of Personality and Social Psychology, 68,* 1995, 261–279. D. I. Perrett, K. A. May, S. Yoshikawa, "Facial shape and judgements of female attractiveness," *Nature, 368,* 1994, 239–242. Zebrowitz et al., 1993.

139 *beautiful Asian, Hispanic, Afro-Caribbean:* Cunningham, Roberts, Barbee, Druen, and Wu, 1995.

139 *beauty in . . . adaptations of beholder:* D. Symons, "Beauty is in the adaptations of the beholder: The evolutionary psychology of human female sexual attractiveness." In P. R. Abramson and S. D. Pinkerton, eds., *Sexual Nature, Sexual Culture* (Chicago: University of Chicago Press, 1995).

140 *As one plastic surgeon said:* Comment made on Discovery Channel/*Discover Magazine* program called "The Science of Sex." Fine Cut Productions, 1996.

140 *Renaissance canons not . . . realistic:* L. G. Farkas, I. R. Munro, and J. C. Kolar, "The validity of Neoclassical facial proportion canons," in L. G. Farkas and I. R. Munro, eds., *Anthropometric Facial Proportions in Medicine* (Springfield: Charles C Thomas, 1987), pp. 57–66.

141 *four hundred attractive faces:* W. Earle Matory, Jr., "Definitions of beauty in the ethnic patient," in Matory, ed., *Ethnic Considerations in Facial Aesthetic Surgery* (Philadelphia: Lippincott-Raven, 1998), pp. 61–83.

141 *golden section:* See M. Ghyka, *The Geometry of Art and Life* (New York: Dover, 1977); and H. E. Huntley, *The Divine Proportion: A Study in Mathematical Beauty* (New York: Dover, 1970).

142 *found in beautiful music and poetry:* A. V. Voloshinov, "Symmetry as a superprinciple of science and art," *Leonardo, 29,* 1996, 109–113.

142 *Fechner's studies:* See C. D. Green, "All that glitters: a review of psychological research on the aesthetics of the golden section," *Perception, 24,* 1995, 937–968.

142 *"numerological fantasies":* M. Gardner, "The cult of the golden ratio," *Skeptical Inquirer, 18,* 1994, 243–247.

142 *proportions:* K. Clark, *The Nude: A Study in Ideal Form* (Princeton: Princeton University Press, 1972), pp. 15, 17.

142 *extensive set of measurements:* R. M. Ricketts, "Divine proportions in facial aesthetics," *Clinics in Plastic Surgery, 9,* 1982, 401–422.

143 *examples of golden sections:* Matory, 1998.

143 *composite photographs:* F. Galton, *Inquiries into Human Faculty and Its Development* (London: Macmillan, 1883).

144 *"founded upon blended memories":* F. Galton, "Generic Images," *Proceedings of the Royal Institution, 9, 1879,* 161–170 (quote from p. 161).

144 *"the special villainous irregularities":* F. Galton, "Composite portraits," *Nature, 18,* 1878, 97–100 (quote from pp. 97–98).

144 *"It is charming":* J. T. Stoddard, "Composite photography," *Century, 33,* 1887, 757.

145 *digitized composites to test the beauty of averages:* J. H. Langlois and L. A. Roggman, "Attractive faces are only average," *Psychological Science, 1,* 1990, 115–121. K Grammer and R. Thornhill, "Human (Homo Sapiens) facial attractiveness and sexual selection: The role of symmetry and averageness," *Journal of Comparative Psychology, 108,* 1994, 233–242. G. Rhodes and T. Tremewan, "Averageness, exaggeration, and facial attractiveness, "*Psychological Science, 7,* 1996, 105–110.

145 *preferences for the average:* J. H. Koeslag, "Koinophilia groups sexual creatures into species, promotes stasis, and stabilizes social behavior," *Journal of Theoretical Biology, 144,* 1990, 15–35.

145 *averageness:* D. Symons, *The Evolution of Human Sexuality* (New York: Oxford, 1981), pp. 195–196.

146 *changes in plastic surgery: American Academy of Facial Plastic and Reconstructive Surgeons Annual Survey,* 1993. Washington, DC.

147 *engagement photographs:* I thank Jeremy Taylor for providing me with Galton's images of couples.

147 *married couples look alike:* V. B. Hinsz, "Facial resemblance in engaged and married couples," *Journal of Social and Personal Relationships, 6,* 1989, 223–229. R. B. Zajonc, P. K. Adelman, S. T. Murphy, and P. M. Niedenthal, "Convergence in the physical appearance of spouses," *Motivation and Emotion, 11,* 1987, 335–346.

147 *husbands and wives are similar:* See J. Diamond, *The Third Chimpanzee: The Evolution and Future of the Human Animal* (New York: HarperCollins, 1992).

147 *studies of Japanese quail:* P. Bateson, "Preferences for cousins in Japanese quail," *Nature, 295,* 1982, 236–237. P. Bateson, "Preferences for close relations in Japanese quail," in H. Ouellet, ed., *Acta XIX Congressus Internationalis Ornithologici. Vol. I* (Ottawa: University of Ottowa Press, 1988), pp. 961–972.

148 *first-cousin pairs:* Bateson, 1982, p. 237. For a fascinating look at human kin attraction, see *Dangerous Reunions*, a film made by Jeremy Taylor.

148 *portraits sometimes bear a resemblance to the artist's face:* See E. H. Gombrich, "The mask and the face: The perception of physiognomic likeness in life and in art," in E. H. Gombrich, J. Hochberg, and M. Black, eds., *Art, Perception, and Reality* (Baltimore: Johns Hopkins Press, 1972), pp. 1–46. For a discussion of the Mona Lisa, see L. Schwartz, *The Computer Artist's Handbook* (New York: W. W. Norton, 1992).

149 *caricature effects:* See G. Rhodes, *Superportraits: Caricatures and Recognition* (East Sussex, UK: Psychology Press, 1996).

150 *extreme traits:* D. I. Perrett, K. A. May, and S. Yoshikawa, "Facial shape and judgements of female attractiveness," *Nature, 368,* 1994, 239–242.

151 *genetic algorithm:* V. S. Johnston and M. Franklin, "Is beauty in the eye of the beholder?" *Ethology and Sociobiology, 13,* 1993, 183–199.

151 *cover girls:* Jones, 1995.

151 *"sex bombs":* R. Dawkins, *River out of Eden: A Darwinian View of Life* (New York: Basic Books, 1995), p. 63.

152 *male and female face differences:* See V. F. Ferrario, C. Sforza, G. Pizzini, G. Vogel, and A. Miani, "Sexual dimorphism in the human face assessed by euclidean matrix analysis," *Journal of Anatomy, 183,* 1993, 593–600. D. H. Enlow, *Handbook of Facial Growth* (Philadelphia: W. B. Saunders, 1982). A. M. Burton, V. Bruce, and N. Dench, "What's the difference between men and women? Evidence from facial measurement," *Perception, 22,* 1993, 153–176.

152 *lips:* K. H. Calhoun, "Lip aesthetics," in K. H. Calhoun and C. M. Steinberg, eds., *Surgery of the Lip* (New York: Thieme, 1992), pp. 11–22.

153 *lower faces:* Johnston and Franklin, 1993.

153 *women apply makeup:* K. Aucoin, *The Art of Makeup* (New York: HarperCollins, 1994), p. 33.

153 *Elizabeth Cady Stanton:* In L. W. Banner, *American Beauty* (Chicago: University of Chicago Press, 1983), p. 49.

153 *Gloria Steinem:* Quoted in M. G. Lord, *Forever Barbie: The Unauthorized Biography of a Real Doll* (New York: William Morrow, 1994), p. 53. Taken from Gloria Steinem, *The Beach Book* (New York: Viking, 1963).

154 *generic image:* I thank Masami Yamaguchi for her observations about women's poses in Japan.

154 *similar images:* M. K. Yamaguchi and M. Oda, "Measuring and creating different facial images for age and gender," *ATR Human Information Processing Research Laboratories Technical Report,* Kyoto, Japan 1996.

154 *baby-faces:* D. S. Berry and L. Z. McArthur, "Perceiving character in faces: The impact of age-related craniofacial changes on social perception," *Psychological Bulletin, 100,* 1986, 3–18.

154 *baby face and attractiveness not the same:* L. A. Zebrowitz and J. M. Montepare, "Impressions of babyfaced individuals across the life span," *Developmental Psychology, 28,* 1992, 1143–1152. K. Atzwanger and

K. Grammer, "Babyness and sexual attraction." Paper available from LBI for Urban Ethology, Vienna, Austria.

155 *dominant and submissive faces:* A. Mazur, J. Mazur, C. Keating, "Military rank attainment of a West Point class: effects of cadets' physical features," *American Journal of Sociology, 90,* 1984, 125–150. U. Mueller and A. Mazur, "Facial dominance of West Point Cadets as a predictor of later military rank," *Social Forces, 74,* 1996, 823–850.

156 *dominant high-school-age boys:* A. Mazur, C. Halpern, and J. R. Udry, "Dominant looking male teenagers copulate earlier," *Ethology and Sociobiology," 15,* 1994, 87–94.

156 *dominance and attractiveness to women:* E. K. Sadalla, D. T. Kenrick, and B. Vershure, "Dominance and heterosexual attraction," *Journal of Personality and Social Psychology, 52,* 1987, 730–738.

156 *masseter muscle:* R. A. Rosa and H. C. Kotkin, "That acquired masseteric look," *Journal of Dentistry for Children,* March-April 1996, pp. 105–7.

156 *bald and baby-faced men:* F. Muscarella and M. R. Cunningham, "The evolutionary significance and social perception of male pattern baldness and facial hair," *Ethology and Sociobiology, 17,* 1996, 99–117.

156 *history of beards:* R. Corson, *Fashions in Hair: The First Five Thousand Years* (London: Peter Owen, 1991).

157 *"swarming with malignant microbes":* From Corson, 1991, p. 571.

157 *Harper's Weekly:* Quoted in ibid., p. 565.

157 *shaved off his mustache:* Otto Fredrick, "When I shaved off my mustache," *New York Times,* November 19, 1982, p. 19. Quoted in Hatfield and Sprecher, 1986, p. 228.

157 *beards have a significant impact on . . . recognition:* R. L. Terry, "Effects of facial transformations on accuracy of recognition," *Journal of Social Psychology, 134,* 1994, 483–492.

157 *Botticelli's Young Man:* Discussed in Lorne Campbell, 1990, p. 12.

158 *"multiple motives":* M. R. Cunningham, A. P. Barbee, C. L. Pike, "What do women want? Facialmetric assessment of multiple motives in the perception of male facial physical attractiveness," *Journal of Personality and Social Psychology, 59,* 1990. 61–72.

158 *"hyper" . . . faces:* T. Hirukawa and M. Yamaguchi, "Effect of sexual dimorphism on human facial attractiveness," *ATR Human Information Processing Research Laboratories technical report,* Kyoto, Japan, 1996.

158 *men and women in Japan and Scotland:* D. I. Perrett, K. J. Lee, I. Penton-Voak, D. Rowland, S. Yoshikawa, D. M. Burt, S. P. Henzi, D. Castles, and S. Akamatsu, "Effects of sexual dimorphism on facial attractiveness," *Nature, 394,* 1998, pp. 884.

159 *smaller smiles:* J. M. Dabbs, Jr., "Testosterone, smiling, and facial appearance," *Journal of Nonverbal Behavior, 21,* 1997, 45–55.

159 *"Minotaur syndrome":* P. G. Morselli, "The Minotaur Syndrome: Plastic surgery of the facial skeleton," *Aesthetic Plastic Surgery, 17,* 1993, 99–102.

159 *Michael Southgate:* Quoted in S. K. Schneider, *Vital Mummies: Performance Design for the Show-window Mannequin* (New Haven: Yale University Press, 1995), p. 87.

160 *"beautiful-handsome-smart-friendly-kind-nice":* From Beloff, 1985, p. 181.

160 *smiles:* K. T. Mueser, B. W. Grau, M. S. Sussman, and A. J. Rosen, "You're only as pretty as you feel: Facial expression as a determinant of physical attractiveness," *Journal of Personality and Social Psychology, 46,* 1984, 469–478. F. M. Deutsch, D. LeBaron, and M. M. Fryer, "What is in a smile?" *Psychology of Women Quarterly, 11,* 1987, 341–352.

160 *pupil size:* E. H. Hess, "Attitude and pupil size," *Scientific American, 212,* 1965, 45–54. M. R. Cunningham, "Measuring the physical in physical attractiveness: Quasi-experiments on the sociobiology of female facial beauty," *Journal of Personality and Social Psychology, 50,* 1986, 925–935.

160 *dilated pupils and research partners:* W. Stass and F. N. Willis, Jr., "Eye contact, pupil dilation, and personal preference," *Psychonomic Science, 7,* 1967, 375–376.

160 *shape of our lips:* D. Morris, *The Naked Ape: a Zoologist's Study of the Human Animal* (New York: McGraw-Hill, 1967).

161 *Alexander Liberman:* Quote from R. T. Lakoff and R. L. Scherr, *Face Value: The Politics of Beauty* (Boston: Routledge and Kegan Paul, 1984), p. 106.

161 *preferred sides:* See I. C. McManus and N. K. Humphrey, "Turning the left cheek," *Nature, 243,* 1973, 272.

161 *fluctuating asymmetries:* S. W. Gangestad, R. Thornhill, and R. A. Yeo, "Facial attractiveness, developmental stability, and fluctuating asymmetry," *Ethology and Sociobiology, 15,* 1994, 73–85. R. Thornhill and S. W. Gangestad, "Human facial beauty: averageness, symmetry and parasite resistance," *Human Nature, 4,* 1993, 237–269. K. Grammer and R. Thornhill, "Human (Homo Sapiens) facial attractiveness and sexual selection: The role of symmetry and averageness," *Journal of Comparative Psychology, 108,* 1994, 233–242.

162 *facial expressions:* Reviewed in N. L. Etcoff, "Asymmetries in recognition of emotion," in F. Boller and J. Grafman, eds., *Handbook of Neuropsychology,* Vol. 3 (Elsevier, 1989), pp. 363–382.

162–63 *facial symmetry and beauty:* G. Rhodes, F. Profitt, J. M. Grady, and A. Sumich, "Facial symmetry and the perception of beauty," *Psychonomic Bulletin and Review,* in press. D. I. Perrett, D. M. Burt, K. J. Lee, D. A. Rowland, and R. E. Edwards, "Fluctuating asymmetry in human faces: symmetry is beautiful," unpublished manuscript, Perception Laboratory, University of St. Andrews, Fife, Scotland. But see also R. Kowner, "Facial asymmetry and attractiveness judgments in developmental perspective," *Journal of Experimental Psychology: Human Perception and Performance, 22,* 1996, 662–675, and J. H. Langlois, L. A. Roggman, and L. Musselman, "What is average and what is not average about attractive faces," *Psychological Science, 5,* 1994, 214–220.

163 *pioneering study:* V. S. Johnston and J. C. Oliver-Rodriguez, "Facial beauty and the late positive component of event-related potentials," *Journal of Sex Research, 34,* 1997, 188–198.

163 *research on stroke patients:* N. L. Etcoff, "Selective attention to facial identity and facial emotion," *Neuropsychologia, 22,* 1984, *281–295.* N. L. Etcoff, "Perceptual and conceptual organization of facial emotions: Hemispheric differences," *Brain and Cognition, 3,* 1984, 385–412.

164 *the right side of the face seems to "resemble" the whole face:* C. Gilbert and P. Bakan, "Visual asymmetry in perception of faces," *Neuropsychologia, 11,* 1973, 355–362.

164 *right hemisphere dominates:* D. M. Burt, and D. I. Perrett, "Perceptual asymmetries in face judgements," *Neuropsychologia, 35,* 1997, 685–693.

164 *brain imaging studies of faces:* H. C. Breiter, N. L. Etcoff, P. J. Whalen, W. A. Kennedy, S. L. Rauch, R. L. Buckner, M. M. Strauss, S. E. Hyman, and B. R. Rosen, "Response and habituation of the human amygdala during visual processing of facial expression," *Neuron, 17,* 1996, 1–13. P. J. Whalen, S. L. Rauch, N. L. Etcoff, S. C. McInerny, M. B. Lee, and M. A. Jenike, "Masked presentations of emotional facial expressions modulate amygdala activity without explicit knowledge," *Journal of Neuroscience, 18,* 1998, 411–418.

165 *brain imaging studies of reward and craving:* See H. C. Breiter, R. L. Gollub, R. M. Weiskoff, D. N. Kennedy, N. Makris, J. D. Berke, J. M. Goodman, H. L. Kanter, D. R. Gastfriend, J. P. Riorden, R. T. Mathew, B. R. Rosen, and S. E. Hyman, "Acute effects of cocaine on human brain activity and emotion," *Neuron, 29,* 1997, 591–611.

165 *prosopagnosics:* Past studies in N. L. Etcoff, R. Freeman, and K. Cave, "Can we lose memories of faces? Content specificity and awareness in a prosopagnosic," *Journal of Cognitive Neuroscience, 3,* 1991, 25–41.

CHAPTER 6: SIZE MATTERS
PAGE

169 *"ornamented by all sorts of combs":* Charles Darwin, *The Descent of Man, and Selection in Relation to Sex* (Princeton: Princeton University Press, 1981), pp. ii, 38.

169 *"strong affections, acute perception":* ibid., pp. ii, 108.

169 *"unarmed, unornamented":* ibid., pp. i, 258.

169 *swordtail fish:* A. Basolo, "Female preference for male sword length in the green swordtail, Xiphorus helleri," *Animal Behavior, 40,* 1990, 332–338.

169 *widowbirds:* M. B. Andersson, "Female choice selects for extreme tail length in a widowbird," *Nature, 299,* 1982, 818–820.

169 *peacock . . . lugs his train:* Helena Cronin, *The Ant and the Peacock* (Cambridge: Cambridge University Press, 1991), p. 185.

169–70 *Peacocks with the most elaborate trains:* M. Petrie, T. Halliday, and C. Sanders, "Peahens prefer peacocks with elaborate trains," *Animal Behavior, 41,* 1991, 323–331.

170   *Great snipes whose white tail patches:* J. Hoglund, M. Eriksson, and L. E. Lindell, "Females of the lek-breeding great snipe, Gallinago media, prefer white tails," *Animal Behavior, 40,* 1990, 23–32.

170   *swallows:* A. P. Moller, "Female choice selects for male sexual tail ornaments in the monogamous swallow," *Nature, 332,* 1988, 640–642.

170   *"Runaway":* R. A. Fisher, *The Genetical Theory of Natural Selection,* 2nd ed. (New York: Dover, 1958).

170   *sustained by popularity alone:* See chapter on explosions and spirals in Richard Dawkins, *The Blind Watchmaker* (New York: W. W. Norton, 1987) pp. 195–220.

170   *handicap principle:* Amotz Zahavi, "Mate selection—A selection for a handicap," *Journal of Theoretical Biology, 53,* 1975, 205–214.

171   *Peacocks with elaborate trains sire offspring:* M. Petrie, "Improved growth and survival of offspring of peacocks with more elaborate trains," *Nature, 371,* 1994, 598–599.

171   *Barn swallows:* A. P. Moller, "Male ornament size as a reliable cue to enhanced offspring viability in the barn swallow," *Proceedings of the National Academy of Science, 91,* 1994, 6929–6932.

171   *Red-throated three-spine sticklebacks:* D. E. Semler, "Some aspects of adaptation in a polymorphism for breeding colours in the Threespine stickleback (Gasterosteus aculeatus)," *Journal of Zoology, 165,* 1971, 291–302.

171   *Flies of the diopsidae family:* G. S. Wilkinson, D. C. Presgraves, and L. Crymes, "Male eye span in stalk-eyed flies indicates genetic quality by meiotic drive suppression," *Nature, 391,* 1998, 276.

171   *tail length and symmetry:* A. P. Moller, "Female swallow preference for symmetrical male sexual ornaments," *Nature, 357,* 1992, 238–239.

172   *reindeer antlers:* E. Markusson and I. Folstad, "Reindeer antlers: Visual indicators of individual quality?" *Oecologia, 110,* 1997, 501–507.

172   *height:* Reviewed in L. F. Martel and H. B. Biller, *Stature and Stigma: The Biopsychosocial Development of Short Males* (Lexington, MA: Lexington Books, 1987). J. S. Gillis, *Too Tall, Too Small.* (Champaign, IL: Institute for Personality and Ability Testing. 1982) R. Keyes, *The Height of Your Life* (Toronto: Little Brown, 1980).

173   *presidential candidates:* J. McGinnis, *The Selling of the President* (New York: Andre Deutsch, 1976).

173   *corporate recruiters:* D. L. Kurtz, "Physical appearance and stature: Important variables in sales recruiting," *Personnel Journal, 48,* 1969, 981–983.

173   *height and income:* I. H. Frieze, J. E. Olson, and D. C. Good, "Perceived and actual discrimination in the salaries of male and female managers," *Journal of Applied Social Psychology, 20,* 1990, 46–67. E. S. Loh, "The economic effects of physical appearance," *Social Science Quarterly, 74,* 1993, 420–438.

173   *no relation to job performance:* W. E. Hensley and R. Cooper, "Height and occupational success: A review and critique," *Psychological Reports, 60,* 1987, 843–849.

174 *"trickle-down theory of torment":* Stephen Hall, "Short like me," *Health,* January/February 1996, p. 98.

174 *"Bond . . . mistrusted":* Ian Fleming, *Goldfinger* (New York: Macmillan, 1959), p. 25.

174 *John Wayne:* From an interview with Robert Mitchum, *Esquire,* February 1983, p. 52.

174 *overestimate own height:* D. J. Dillon, "Measurement of perceived body size," *Perceptual and Motor Skills, 14,* 1962, 191–196.

174 *unconscious association of power, status, and height:* P. R. Wilson, "Perceptual distortion of height as a function of ascribed academic status," *Journal of Social Psychology, 74,* 1968, 97–102. W. D. Dannenmaier and F. J. Thumin, "Authority status as a factor in perceptual distortion of size," *Journal of Social Psychology, 63,* 1964, 361–365.

175 *personal ads:* M. Lynn and B. A. Shurgot, "Responses to lonely hearts advertisements: Effects of reported physical attractiveness, physique and coloration," *Personality and Social Psychology Bulletin, 10,* 1984, 349–357. See also Linda A. Jackson, *Physical Appearance and Gender: Sociobiological and Sociocultural Perspectives* (Albany: State University of New York Press, 1992).

175 *height stereotypes:* L. A. Jackson and K. S. Ervin, "Height stereotypes of women and men: The liability of shortness for both sexes," *Journal of Social Psychology, 132,* 1991, 433–445.

175 *in one study women preferred men:* W. Graziano, T. Brothen, and E. Berscheid, "Height and attraction: Do men see women eye to eye?" *Journal of Personality, 46,* 1978, 128–145.

175 *Accra immigrants:* Tim Sullivan, "Ghana's tall men suffer," posted February 27, 1997 by Canadian Free Radio Association.

175 *married couples:* J. S. Gillis and W. E. Avis, "The male-taller norm in mate selection," *Personality and Social Psychology Bulletin, 6,* 1980, 396–401.

175 *potential sperm donors:* J. E. Scheib, "Women's choices of sperm donors: So many donors, so little information." Paper presented at the *Human Behavior and Evolution Society Eighth Annual Conference,* Northwestern University, Evanston, Illinois, June 26–30, 1996.

176 *half of the women wanted to be shorter:* G. Calden, R. M. Lundy, and R. S. Schlafer, "Sex differences in body concepts," *Journal of Consulting Psychology, 23,* 1959, 378.

177 *Studies that focus on weight:* A. E. Fallon and P. Rozin, "Sex differences in perceptions of desirable body shape," *Journal of Abnormal Psychology, 94,* 1985, 102–105. P. Rozin and A. Fallon, "Body image, attitudes to weight, and misperceptions of figure preferences of the opposite sex: A comparison of men and women in two generations," *Journal of Abnormal Psychology, 97,* 1988, 342–345. S. E. Beren, H. A. Hayden, D. E. Wilfley, and C. M. Grilo, "The influence of sexual orientation on body dissatisfaction in adult men and women," *International Journal of Eating Disorders, 20,* 1996, 135–141. A. Feingold and R. Mazzela, "Gender differences in body image are increasing," *Psychological Science, 9,* 1998, 190–195.

177   *Many men consider themselves underweight:* L. B. Mintz and N. E. Betz, "Sex differences in the nature, realism, and correlates of body image," *Sex Roles, 15,* 1986, 185–195. C. Davis and M. P. Cowles, "Body image and exercise: A study of relationships and comparisons between physically active men and women," *Sex Roles, 25,* 1991, 33–44.

177   *preference for V-shaped torso:* P. J. Lavrakas, "Female preference for male physiques," *Journal of Research in Personality, 9,* 1975, 324–334. R. A. Maier and P. J. Lavrakas, "Attitudes toward women, personality rigidity, and idealized physique preferences in males," *Sex Roles, 11,* 1984, 425–433. T. Horvath, "Correlates of physical beauty in men and women," *Social Behavior and Personality, 7,* 1979, 145–151. C. Davis, H. Brewer, and M. Weinstein, "A study of appearance anxiety in young men," *Social Behavior and Personality, 21,* 1993, 63–74.

177   *Caucasian and Japanese male students:* A. Arkoff and B. Weaver, "Body image and body dissatisfaction in Japanese Americans," *Journal of Social Psychology, 68,* 1966, 323–330.

177–78   *sex differences in strength:* From Mary Anne Baker, ed., *Sex Differences in Human Performance* (New York: John Wiley, 1987).

178   *pectoral implants, calf muscle implants:* Amy M. Spindler, "It's a face-lifted, tummy-tucked jungle out there," *New York Times,* June 9, 1996, Sec. 3, pp. 1, 8, 9.

179   *Arnold Schwarzenegger:* See K. R. Dutton, *The Perfectible Body: The Western Ideal of Physical Development* (London: Cassell, 1995).

179   *Sylvester Stallone:* Quote from Susan Faludi, "The masculine mystique," *Esquire,* December 1996, p. 91.

179   *"If fat is a feminist issue":* Quote by John Webb, cited in Dutton, 1995, p. 259.

179   *muscle dysmorphia:* H. G. Pope, D. L. Katz, and J. I. Hudson, "Anorexia Nervosa and 'Reverse Anorexia' among 108 male bodybuilders," *Comprehensive Psychiatry, 34,* 1993, 406–409. H. G. Pope, A. J. Gruber, R. Olivardia, and K. A. Phillips, "Muscle dysmorphia: An underrecognized form of body dysmorphic disorder," *Psychosomatics, 38,* 1997, 548–557.

180   *steroid use in high school:* W. A. Buckley, C. E. Yesalis, K. E. Friedl, W. Anderson, A. Streit, and J. Wright, "Estimated prevalence of anabolic steroid use among male high school seniors," *Journal of the American Medical Association, 260,* 1988, 3441–3445. R. H. DuRant, V. I. Rickert, C. S. Ashworth, C. Newman, and G. Slavens, "Use of multiple drugs among adolescents who use anabolic steroids," *New England Journal of Medicine, 328,* 1993, 922–926.

180   *mannequins imported from Europe:* In S. K. Schneider, *Vital Mummies: Performance Design For the Show-window Mannequin* (New Haven: Yale University Press, 1995).

180   *body dysmorphic disorder:* K. A. Phillips, *The Broken Mirror: Understanding and Treating Body Dysmorphic Disorder* (New York: Oxford University Press, 1996), p. 68.

181 *A recent sample of over a thousand men:* J. Sparling, "Penile erections: shape, angle, and length," *Journal of Sex and Marital Therapy, 23,* 1997, 195–207.

181 *proportionately largest penis:* See W. G. Eberhard, *Sexual Selection and Animal Genitalia* (Cambridge: Harvard University Press, 1985). R. V. Short, "Testis weight, body weight, and breeding systems in primates," *Nature, 293,* 1981, 55. M. Kirkpatrick, "Sexual selection: Is bigger always better," *Nature, 337,* 1989, 116.

181 *surgery to lengthen or widen their penises:* R. H. Stubbs, "Penis lengthening—A retrospective review of 300 consecutive cases," *Canadian Journal of Plastic Surgery, 5,* 1997, 93–100.

182 *"locker room phobia":* ibid.

182 *males present . . . genitals:* Irenaus Eibl-Eibesfeldt, *Love and Hate* (New York: Holt, Rinehart and Winston, 1971).

182 *never regard . . . genitals . . . as beautiful:* Sigmund Freud, *Three Essays on the Theory of Sexuality* (New York: Basic Books, 1962), p. 22.

182 *her lover's genitals:* Sylvia Plath, *The Bell Jar* (New York: Alfred Knopf, 1963).

183 *penises risk "ludicrousness":* Camille Paglia, *Sexual Personnae: Art and Decadence from Nefertiti to Emily Dickinson* (New York: Vintage, 1991), p. 17.

183 *"internal courtship devices":* W. Eberhard, 1985.

183 *the standard thinking in sex research:* W. H. Masters and V. E. Johnson, *Human Sexual Response* (Boston: Little Brown, 1966). See also W. A. Fisher, N. R. Branscombe, and C. R. Lemery, "The bigger the better? Arousal and attributional responses to erotic stimuli that depict different size penises," *Journal of Sex Research, 19,* 1983, 377–396.

184 *penises can come equipped:* J. E. Lloyd, "Firefly communication," *Anima,* June 1979, p. 32.

184 *origin of genital extravagance:* G. Arnqvist, "Comparative evidence for the evolution of genitalia by sexual selection," *Nature, 393,* 1998, 784–786. D. T. Gwynne, "Genitally does it," *Nature 393,* 1998, 734–735.

185 *facial attractiveness and symmetry:* S. W. Gangestad, R. Thornhill, and R. A. Yeo, "Facial attractiveness, developmental stability, and fluctuating asymmetry," *Ethology and Sociobiology, 15,* 1994, 73–85.

185 *body mass and symmetry:* J. T. Manning. "Fluctuating asymmetry and body weight in men and women: Implications for sexual selection," *Ethology and Sociobiology, 16,* 1995, 145–153.

185 *symmetrical scorpionflies emit preferred pheromones:* R. Thornhill, "Female preference of males with low fluctuating asymmetry in the Japanese scorpionfly (Panorpa japonica: Mecoptera)," *Behavioral Ecology, 3,* 1992, 277–283.

185 *symmetrical flowers:* A. P. Moller and M. Eriksson, "Patterns of fluctuating asymmetry in flowers: Implications for sexual selection in plants," *Journal of Evolutionary Biology, 7,* 1994, 97–113. A. P. Moller, "Bumblebee pref-

erence for symmetrical flowers," *Proceedings of the National Academy of Science, 92,* 1995, 2288–2292.

185   *symmetry . . . as a measure for overall fitness:* See R. Thornhill and A. P. Moller, "Developmental stability, disease, and medicine," *Biological Reviews of the Cambridge Philosophical Society, 72,* 1997, 497–548.

185   *review of sixty-two studies of forty-one species:* A. P. Moller and R. Thornhill, "Bilateral symmetry and sexual selection: A meta-analysis," *American Naturalist, 151,* 1998, 174–192.

186   *symmetry and number of partners:* R. Thornhill and S. W. Gangestad, "Human fluctuating asymmetry and sexual behavior," *Psychological Science, 5,* 1994, 297–302.

186   *symmetry and orgasm:* R. Thornhill, S. W. Gangestad, and R. Comer, "Human female orgasm and mate fluctuating asymmetry," *Animal Behavior, 50,* 1995, 1601–1615.

186   *breast asymmetry:* A. P. Moller, M. Soler, and R. Thornhill, "Breast asymmetry, sexual selection, and human reproductive success," *Ethology and Sociobiology, 15,* 1995, 207–219.

186–87   *symmetry and ovulation:* D. Scutt and J. T. Manning, "Symmetry and ovulation in women," *Human Reproduction, 11,* 1996, 2477–2480.

187   *breasts:* For an insightful history of how the breast has been perceived in the Western world, see Marilyn Yalom, *A History of the Breast* (New York: Alfred Knopf, 1997).

187   *"visitor of another species":* John Steinbeck, *The Wayward Bus* (New York: Viking Press, 1947), p. 5.

187   *breasts as deceptive:* B. S. Low, R. D. Alexander, K. M. Noonan, "Human hips, breasts and buttocks: Is fat deceptive?" *Ethology and Sociobiology, 8,* 1986, 249–257.

187   *review of hypotheses about breasts:* See T. M. Caro, "Human breasts: Unsupported hypotheses," *Human Evolution, 2,* 1987, 271–282. T. M. Caro and D. W. Sellen, "The reproductive advantages of fat in women," *Ethology and Sociobiology, 11,* 1990, 51–66.

188   *sports bras:* "The turbulent world of swimming (designing swimsuits that reduce drag that comes from having breasts)," *Economist,* June 7, 1997, p. 82.

188   *breasts to shift the male interest to the front:* See Desmond Morris, *The Naked Ape* (New York: McGraw Hill, 1967).

188   *breasts sexy only in young women:* Donald Symons, personal communication.

189   *implants:* See Geoffrey Cowley, "Silicone: Juries vs. Science," *Newsweek,* November 13, 1995, p. 75. Denise Grady, "Cosmetic breast enlargements are making a comeback," *New York Times,* July 21, 1998, p. C7.

191   *sex differences in WHR:* M. Rebuffe-Scrive, "Regional adipose tissue metabolism in men and in women during menstrual cycle, pregnancy, lactation, and menopause," *International Journal of Obesity, 11,* 1987, 347–355. Y. Tahara, N. Tsunawake, K. Yukawa, N. Yamaski, K. Nishiyama, H. Urata, K. Katsuno, and Y. Fukuyama, "Sex differences in interrelation-

ships between percent body fat (%fat) and waist-to-hip ratio (WHR) in healthy male and female adults," *Annals of Physiological Anthropology*, *13*, 1994, 293–301.

191 *women with polycystic ovary disease:* D. J. Evans, J. H. Barth, and C. W. Burke, "Body fat topography in women with androgen excess," *International Journal of Obesity*, *12*, 1988, 157–162. M. Rebuffe-Scrive, G. Culberg, P. A. Lundberg, G. Lindstedt, and P. Bjorntorp, "Anthropometric variables and metabolism in polycystic ovarian disease," *Hormone Metabolic Research*, *21*, 1989, 391–397.

191–92 *fertility and WHR:* B. M. Zaadstra, J. C. Seidell, P. A. H. Van Noord, E. R. te Velde, J. D. F. Habbema, B. Vrieswijk, and J. Karbaat, "Fat and female fecundity: Prospective study of effect of body fat distribution on conception rates," *British Medical Journal*, *306*, 1993, 484–487. P. Wass, U. Waldenstron, and D. Hellberg, "An android body fat distribution in females impairs the pregnancy rate of in-vitro fertilization-embryo transfer," *Human Reproduction*, *12*, 1997, 2057–2060.

192 *Devendra Singh's studies:* D. Singh, "Adaptive significance of female physical attractiveness: Role of waist-to-hip ratio," *Journal of Personality and Social Psychology*, *65*, 1993, 293–307. D. Singh, "Female judgment of male attractiveness and desirability for relationships: Role of waist-to-hip ratio and financial status," *Journal of Personality and Social Psychology*, *69*, 1995, 1089. D. Singh and S. Luis, "Ethnic and gender consensus for the effect of waist-to-hip ratio on judgment of women's attractiveness," *Human Nature*, *6*, 1995, 51–65. D. Singh, "Body fat distribution and perception of desirable body shape by young black men and women," *International Journal of Eating Disorders*, *16*, 1994, 289–294.

193 *supermodel measures:* M. J. Tovee, S. M. Mason, J. L. Emery, S. E. McClusky, and E. M. Cohen-Tovee, "Supermodels: stick insects or hourglasses?" *Lancet*, *350*, 1997, 1474–1475.

194 *WHR and health:* P. Bjorntorp, "The association between obesity, adipose tissue distribution and disease," *Acta Medica Scandanvica (Suppl. J)*, *723*, 1988, 121–134. J. Tichet, S. Vol, B. Balkau, H. Le Clesiau, and A. D'Hour, "Android fat distribution by age and sex: The waist hip ratio," *Diabetes Metabolism*, *19*, 1993, 273–276.

194 *obesity has such a great influence:* D. Singh and R. K. Young, "Body weight, waist-to-hip ratio, breasts, and hips: Role in judgments of female attractiveness and desirability for relationships," *Ethology and Sociobiology*, *16*, 1995, 483–507.

194 *fashion and the waist:* See, for example, Suzy Menkes, "The hourglass again: Remember Mae West," *New York Times*, March 5, 1995, pp. 37, 41.

195 *Veronica Webb:* From Jody Shields, "Arch enemies, the dangerous lure of high heels," *Mirabella*, September 1994.

195 *Manolo Blahnik:* In Hilary Sterne, "Standing Tall," *Elle*, July 1997, p. 86.

196 *forty billion dollars a year:* See Jay Palmer, "Hey fatso! Despite a glut of diet foods and health clubs, Americans are growing plumper," *Barrons*, July 1, 1996, pp. 25–29.

196 *number of obese in the United States:* Marian Burros, "Despite awareness of risks more in U.S. are getting fat," *New York Times*, July 17, 1994, pp. 1,

23. "Most Americans are overweight," *New York Times,* November 20, 1996, pp. C1, C3.

196   *obesity in Britain:* Georgina Ferry, "The Fat of the Land," *New Scientist, 22,* April 1995, p. 26.

196   *Fat consumption dropped from thirty-six percent to thirty-four percent:* Jane E. Brody, "Why bad health habits drive out good ones," *New York Times,* February 1, 1995, p. C9.

197   *average American spends . . . sixteen minutes a day exercising:* Geoffrey Cowley with Karen Springen, "Critical Mass," *Newsweek,* September 25, 1995, pp. 66–67.

197   *history of fat consumption:* S. M. Garn, "From the Miocene to olestra: A historical perspective on fat consumption," *Journal of the American Dietary Association, 97,* 1997, S54–57.

197   *almost half of developing societies have food shortages:* P. J. Brown and M. Konner, "An anthropological perspective on obesity," *Annals of the New York Academy of Sciences, 499,* 1987, 29–46.

197   *hearts of palm:* Napoleon A. Chagnon, *Yanomamo: The Last Days of Eden* (San Diego: Harcourt Brace Jovanovich, 1992).

198   *Tonga:* See Elizabeth MacLean, *The 1986 National Nutrition Survey of the Kingdom of Tonga: Summary Report Prepared for the National Food and Nutrition Committee* (New Caledonia: South Pacific Commission, 1992).

199   *"fattening rooms":* P. J. Brink, "The fattening room among the Annang of Nigeria," *Medical Anthropology, 12,* 1989, 131–143.

199   *studies in Fiji:* Anne E. Becker, *Body, Self, and Society: The View from Fiji* (Philadelphia: University of Pennsylvania Press, 1995).

199   *studies in Britain, Uganda, and Kenya:* A. Furnham and N. Alibhai, "Cross-cultural differences in the perception of female body shapes," *Psychological Medicine, 13,* 1983, 829–837. A. Furnham and P. Baguma, "Cross-cultural differences in the evaluation of male and female body shapes," *International Journal of Eating Disorders, 15,* 1994, 81–89.

200   *prestige . . . conferred by signs of abundance:* M. Mackenzie, "The Pursuit of slenderness and addiction to self-control," in Jean Weininger and George M. Briggs, eds., *Nutrition Update.* Vol. 2 (New York: John Wiley and Sons, 1985).

200   *obesity and SES:* J. Sobal and A. J. Stunkard, "Socioeconomic status and obesity: A review of the literature," *Psychological Bulletin, 105,* 1989, 260–275.

200   *thin women . . . "marry up":* Reviewed in ibid. See also S. M. Garn, T. V. Sullivan, and V. M. Hawthorne, "Educational level, fatness, and fatness differences between husbands and wives," *American Journal of Clinical Nutrition, 50,* 1989, 740–745.

201   *genetics and weight:* Reviewed in Sobal and Stunkard, 1989.

201   *Miss Americas:* See A. Mazur, "U.S. trends in feminine beauty and overadaptation," *Journal of Sex Research, 22,* 1986, 281–303.

201 *"an epidemic illness"*: Hilde Bruch, *The Golden Cage: The Enigma of Anorexia Nervosa* (Cambridge: Harvard University Press, 1978), pp. vii–viii.

202 *year of bulimia*: *Newsweek*, January 4, 1981, p. 26.

202 *"easy to become an anorexic"*: Naomi Wolf, *The Beauty Myth: How Images Are Used Against Women* (New York: Anchor, 1992), pp. 201, 208.

202 *eating disorders . . . infrequent*: Joan Jacobs Brumberg, *Fasting Girls: The History of Anorexia* (New York: Plume, 1989).

202 *prevalence of eating disorders*: K. J. Hart and T. H. Ollendick, "Prevalence of bulimia in working and university women," *American Journal of Psychiatry, 142*, 1985, 851–854. A. Drewnowski, D. K. Lee, and D. D. Kahn, "Bulimia in college women: Incidence and recovery rates," *American Journal of Psychiatry, 145*, 1988, 753–755. E. Fombonne, "Anorexia nervosa: No evidence of an increase," *British Journal of Psychiatry, 166*, 1995, 462–471. "Eating Disorders: Recent Advances," Symposium held at the American Psychiatric Association, May 20, 1995, summarized in R. Pies, "New directions in the diagnosis and treatment of eating disorders," *Advances in Psychiatric Medicine*, October 1995.

203 *women with eating disorders suffer disruptions in fertility*: D. E. Stewart, E. Robinson, D. S. Goldbloom, and C. Wright, "Infertility and eating disorders," *American Journal of Obstetrics and Gynecology, 163*, 1990, 1196–1199. D. E. Stewart, "Reproductive functions in eating disorders," *Annals of Medicine, 24*, 1992, 287–291.

203 *severe dietary restriction in animals*: See J. F. Nelson, K. Karelus, M. D. Bergman, L. S. Felicio, "Neuroendocrine involvement in aging: Evidence from studies of reproductive aging and caloric restriction," *Neurobiology of Aging, 16*, 1995, 837–843. S. Austad, "Aging and caloric restrictions: Human effects and mode of action," *Neurobiology of Aging, 16*, 1995, 851–852.

203 *nonconscious strategy to control reproduction*: For example, see M. T. McGuire, *Darwinian Psychiatry* (New York: Oxford University Press, 1998).

CHAPTER 7: FASHION RUNAWAY
PAGE

207 *"vagaries of fashion"*: Ovid, *Art of Love* (Bloomington, IN: Indiana University Press, 1957), p. 157.

207 *fashion like fruit flies*: Quentin Bell, *On Human Finery*, 2nd ed. (New York: Schocken, 1976).

208 *clothes like a language*: See, for example, Alison Lurie, *The Language of Clothes* (New York: Random House, 1981).

208 *clothes are talking*: Ibid., p. 261.

209 *Katherine Hamnett*: Quote from Paul Mather, "Hamnett," *Blitz*, November 1989, p. 27.

209 *Tom Ford*: Quote from Suzy Menkes, "Sex and single guys strut into the limelight: a youthful look with attitude," *International Herald Tribune*, January 25, 1997, p. 19.

209   *Sexuality in religious art:* See Leo Steinberg, *The Sexuality of Christ in Renaissance Art and in Modern Oblivion* (New York: Pantheon, 1983). For an interesting discussion of related issues, see John Updike, "Can genitals be beautiful?" *New York Review of Books,* December 4, 1977, pp. 10–12.

209   *clothes add sex appeal:* Anatole France, *Penguin Island.* Trans. A. W. Evans (Norwalk, CT: Heritage Press, 1975), p. 41.

210   *crinoline . . . "a hollow sham":* James Laver, *Costume and Fashion: A Concise History* (New York: Thames and Hudson, 1985), p. 184.

210   *women liberated from corsets:* Paul Poiret, from *En habillant l'époque,* Paris. 1930, pp. 136–137. Trans. and quoted in Bell, 1976, p. 53.

211   *Condé Nast bought Vogue:* Kennedy Fraser, "Introduction," in A. Liberman and A. Wintour, eds., *On the Edge: Images from 100 years of Vogue* (New York: Random House, 1992), pp. 2, 3.

212   *chimps:* See Frans de Waal, *Good Natured: The Origins of Right and Wrong in Humans and Other Animals* (Cambridge: Harvard University Press, 1996).

212   *Russian graves:* See H. Knecht, A. Pike-Tay, and R. White, *Before Lascaux: The Complex Record of the Early Upper Paleolithic* (Boca Raton: CRC Press, 1993). See also Steven Mithen, *The Prehistory of the Mind: The Cognitive Origins of Art, Religion, and Science* (London: Thames and Hudson, 1996).

213   *Fourteenth-century fashions:* In B. Payne, G. Winakor, and J. Farrell-Beck, *The History of Costume,* 2nd ed. (New York: Harper Collins, 1992). Note also that fashion was influenced by Crusades of 1095–1291, which had exposed Europeans to the beauty of the fabrics of the Orient. The plague had killed thirty to forty percent of Europe's population, and the survivors were encouraged to turn away from mourning and celebrate their survival. In 1348 in Venice, the Senate forbade anyone to show signs of mourning without incurring heavy fines. The aim of the law was to induce "great joy and gaiety" for having survived danger and rid the citizens of black mourning clothes. Paolo Selmi, "Fashion and luxuries in the political mentality of the Venetian Republic," in *I Mestieri della Moda a Venezia Serenissima: The Arts of Fashion in Venice from the 13th to the 18th Century.* Curator, Doretta Davanza Poli. Catalogue to exhibition at The Equitable Gallery, New York City, 1995.

214   *middle classes . . . voracious consumers:* S. Bayley, *Taste: The Secret Meaning of Things* (New York: Pantheon, 1991), p. 27.

214   *consuming:* J. B. Schor, *The Overspent American: Upscaling, Downshifting, and the New Consumer* (New York: Basic Books, 1998), p. 6.

214   *clothes . . . establish social position:* Thorstein Veblen, *The Theory of the Leisure Class* (New York: Penguin, 1994) p. 36.

215   *"patagonia couture":* "Sweat Equity," *Mirabella,* March/April 1998. p. 74.

215   *importance of war and sport:* Bell, 1976, p. 43.

215   *high heels:* Jennifer Steinhauer, "Walk softly and make a big statement," *New York Times,* May 17, 1998, Sec. 9, pp. 1, 2.

216   *qualities of the courtier:* Baldassare Castiglione, *The Courtier* (New York: Penguin, 1976).

216   *Louis XIV:* From Vincent Cronin, *Louis XIV* (Boston: Houghton Mifflin, 1965). Norbert Elias, *The Court Society* (Oxford: Basil Blackwell, 1983).

217   *Stephen Tennant:* Philip Hoare, *Serious Pleasures: The Life of Stephen Tennant* (New York: Hamish Hamilton, 1991).

217   *Ottoline Morrell:* From Miranda Seymour, *Ottoline Morrell: Life on the Grand Scale* (London: Sceptre, 1993), p. 99.

217   *muses:* See Dominick Dunne, "Paris when it sizzles," *Vanity Fair*, May 1998, pp. 198–204, 246–250.

218   *Venetian sumptuary laws:* In Selmi, 1995.

218   *minimalist aesthetic known as iki:* Liza Dalby, *Kimono: Fashioning Culture.* (New Haven: Yale University Press, 1993).

219   *poulaines:* François Boucher, *20,000 Years of Fashion* (New York: Harry N. Abrams, 1987). Payne, Winakor, and Farrell-Beck, 1992.

219   *Venetian Episcopal synod:* Selmi, 1988, p. 39.

219   *status symbols:* Erving Goffman, "Symbols of class status," *British Journal of Sociology, 2,* 1951, 294–304.

220   *democratization of fashion:* See Philippe Perrot, *Fashioning the Bourgeoisie,* trans. R. Bienvenu (Princeton: Princeton University Press, 1994). Gilles Lipovetsky, *The Empire of Fashion,* trans. Catherine Porter (Princeton: Princeton University Press, 1994).

221   *haute couture:* Skrebneski, *The Art of Haute Couture.* Text by Laura Jacobs (New York: Abbeville Press, 1995).

221   *Vionnet . . . an artist:* Bruce Chatwin, *What Am I Doing Here?* (New York: Penguin, 1989), pp. 86–87.

221   *no more than three thousand women:* Dunne, 1998, pp. 204, 200.

222   *Hilfiger:* In J. Schor, 1998, p. 46.

222   *Donna Karan:* In Katherine Betts, "Rumblings in the ranks," *Vogue*, April 1998, p. 322.

222   *twenty-five percent of designer sunglasses:* J. Schor, 1998, p. 56.

223   *Palm Springs:* Ann Japenga, "Face lift city," *Health*, March-April 1993, pp. 46–55.

224–25   *model paid to advertise clothing:* Marcus Schenkenberg, *New Rules* (New York: Universe, 1997), p. 48.

225   *modeling:* Robert Frank and Philip J. Cook, *The Winner-Take-All Society* (New York: Penguin, 1995).

226   *the Sun has banned: The Face*, May 1998, p. 9.

226   *"style surfing," signifying references:* Ted Polhemus, *style surfing* (London: Thames and Hudson, 1996), pp. 12, 15.

227   *Air max metallics:* Peter Lyle and Laura Craik, "Herd Times," *The Face*, May 1998, p. 199.

228   *Betsey Johnson:* In K. Hamilton and K. Martineau, "Future fashions are not quite ready to wear," *Newsweek*, August 4, 1997, p. 11.

228   *smart clothes:* Alex. P. Pentland, "Smart Rooms, Smart Clothes," *Scientific American*, April 1998, p. 124. Steve Mann, "Smart clothing: The wearable computer and wearcam," *Personal Technologies*, *1*, March 1997. "Computer Couture," *Vogue*, February 1998, p. 124.

CHAPTER 8: CONCLUSION
PAGE

233   *experience of beauty:* Marvin Minsky, "Negative Expertise," *International Journal of Expert Systems*, 7, 1994, pp. 13–19.

234   Peter Schjeldahl, "Notes on beauty," in Bill Beckley with David Shapiro, eds., *Uncontrollable Beauty* (New York: Allworth Press, 1998), p. 58.

234   *"Scarlett O'Hara was not beautiful":* Margaret Mitchell, *Gone With the Wind* (New York: Scribners, 1964), p. 1.

234   *courtship signals:* in M. M. Moore, "Nonverbal courtship patterns in women: Context and consequences," *Ethology and Sociobiology*, 6, 237–247; and M. M. Moore and D. L. Butler, "Predictive aspects of nonverbal courtship behavior in women," *Semiotica*, 76, 205–215.

235   *Letterman anecdote:* Elizabeth Weil, "Who is Barry White and why do women love him?" *Boston Phoenix*, July 7–13, 1995, pp. 4–5.

235   *animal expressions of love:* Charles Darwin, *The Descent of Man, and Selection in Relation to Sex* (Princeton: Princeton University Press, 1981), pp. ii, 330.

236   *the male proboscis monkey:* See Helena Cronin, *The Ant and the Peacock* (Cambridge: Cambridge University Press, 1993), p. 176.

236   *Vocal attractiveness:* M. Zuckerman and R. E. Driver, "What sounds beautiful is good: The vocal attractiveness stereotype," *Journal of Nonverbal Behavior*, *13*, 1989, 67–82. M. Zuckerman, K. Miyake, and H. S. Hodgins, "Cross-channel effects of vocal and physical attractiveness and their implications for interpersonal perception," *Journal of Personality and Social Psychology*, *60*, 1991, 545–554.

236   *Male and female voices:* David Graddol and Joan Swann, *Gender Voices* (Oxford: Basil Blackwell, 1989).

237   *Margaret Thatcher changes pitch:* Graddol and Swann. *Supermodel voices and changes in elite accent* from an interview with Sam Chwat, M.S., CCC-SP, director of New York Speech Improvement Services, on July 28, 1995. See also J. Jennifer Steinhauer's article on Chwat: "New York: Where ethnic is out and speech therapy is in," *International Herald Tribune* (August 12, 1993), pp 1–2.

237   Charles Baudelaire, *The Flowers of Evil*, trans. M. and J. Mathews (New York: New Directions, 1989), p. 32.

238   Brian Eno, "Scents and Sensibility," *Details*, July 1992.

238   *Human VNO:* L. Monti-Block, C. Jennings-White, D. S. Dolberg, and D. L. Berliner, "The human vomeronasal system," *Psychoneuroendocrinology*, *19*, 1994, 673–686.

238 *women synchronizing:* K. Stern and M. K. McClintock, "Regulation of ovulation by human pheromones," *Nature, 392,* 1998, 177–179. See also A. Weller, "Communication through body odor," *Nature, 392,* 1998, 126–127.

239 *Women find androstenones neutral at ovulation:* See Karl Grammer, "5-a-androst-162n-3a-on: A male pheromone? A brief report," *Ethology and Sociobiology, 14,* 1993, 201–207.

239 *Women and men exposed to androstenol judging photographs:* See M. D. Kirk-Smith, D. A. Booth, D. Carroll, and P. Davies, "Human social attitudes affected by androstenol," *Research Communications in Psychology, Psychiatry, and Behavior, 3,* 1978, 379–384. *Women's moods affected by androstenol:* See D. Benton, "The influence of androstenol—a putative human pheromone—on mood throughout the menstrual cycle," *Biological Psychology, 15,* 1982, 249–256. *Androstenone sprayed on chairs:* See M. D. Kirk-Smith and D. A. Booth, "Effects of androstenone on choice of location in others' presence." In H. van der Starre, ed., *Olfaction and Taste,* Vol. 7 (London: IRL Press, 1980).

239 *perfumes:* See D. M. Stoddardt, *The Scented Ape: The Biology and Culture of Human Odor* (Cambridge: Cambridge University Press, 1991). This is also an excellent general reference on odor. For another, see P. Vroon, *Smell: The Secret Seducer* (New York: Farrar, Straus, and Giroux, 1994).

240 *Peppermint research:* P. N. Badia, N. Wesensten, W. Lammers, J. Culpeper, and J. Harsh, "Responsiveness to olfactory stimuli presented in sleep," *Physiology and Behavior, 48,* 1990, 87–90.

240 *Lemon odors reseach:* Teruhisa Komori, Ryoichi Fujiwara, Masahiro Tanida, Junichi Nomura, and Mitchel M. Yokoyama, "Effects of citrus fragrance on immune function and depressive states," *Neuroimmunomodulation, 2,* 1995, 174–180.

240 *Female keener sensitivity:* R. L. Doty, L. S. Applebaum, H. Zusho, and R. G. Settle, "Sex difference in odor identification ability: a cross-cultural analysis," *Neuropsychologia, 23,* 1985, 667–672. R. L. Doty, P. J. Snyder, G. R. Huggins, and L. D. Lowry, "Endocrine, cardiovascular and psychological correlates of olfactory sensitivity changes during the human menstrual cycle," *Journal of Comparative Physiology and Psychology, 95,* 1981, 45–51.

240 *MHC genes:* C. Wedekind, T. Seebeck, F. Bettens, and Al J. Paepke, "MHC-dependent mate preferences in humans," *Proceedings of the Royal Society of London,* B260, 1995, 245–249. C. Wedekind and S. Furi, "Body odour preferences in men and women: do they aim for specific MHC combinations or simple heterozygosity?" *Proceedings of the Royal Society of London,* B264, 1997, 1471–1479.

240 *asked . . . men to wear T-shirts:* Wedekind and Furi, p. 1479.

241 *beauty is . . . unfair:* Tom Wolfe, "Funky Chic," in Jann Wenner, ed., *20 Years of Rolling Stone: What a Long, Strange Trip It's Been* (New York: Friendly Press, 1987), p. 212.

242 *have to understand beauty:* Lester Bangs, "The noise supremacists" in Greil Marcus, ed., *Psychotic Reactions and Carburetor Dung* (New York: Vintage Books, 1988), p. 282.

242 *beautiful not necessarily good:* Roger Brown: *Social Psychology*, 2nd ed. (New York: Free Press, 1986, pp. 395–396.

243 *"allowing beautiful women their beauty":* Karen Lehrman, *The Lipstick Proviso* (New York: Anchor, 1997).

244 *pleasure of beauty:* Bertrand Russell, *The Conquest of Happiness* (New York: Liveright, 1958), p. 90.

244 *life of George Eliot:* G. S. Haight, *George Eliot: A Biography* (London: Penguin, 1968). *Clematis quote:* From G. S. Haight, *The George Eliot Letters* (London: Oxford University Press, 1954). For an interesting discussion of women writers of the eighteenth and nineteenth centuries and beauty, see Ellen Zetzel Lambert, *The Face of Love: Feminism and the Beauty Question* (Boston: Beacon Press, 1995).

245 *Henry James's letter to his father, May 10, 1869:* In Leon Edel, ed., Henry James *Selected Letters* (Cambridge: Harvard University Press, 1987), p. 35.

245 *George Eliot quote: Adam Bede* (New York: Penguin, 1981), p. 177.

References

Abbey, A. Sex differences in attributions for friendly behavior. Do males misperceive females' friendliness? *Journal of Personality and Social Psychology, 42,* 1982, 830–838.

Anderson, R. Physical attractiveness and locus of control. *Journal of Social Psychology, 105,* 1978, 213–216.

Andersson, M. B. Female choice selects for extreme tail length in a widowbird. *Nature, 299,* 1983, 818–820.

Appleton, J. *The Experience of Landscape.* New York: John Wiley, 1975.

Arkoff, A. & Weaver, N. Body image and body dissatisfaction in Japanese Americans. *Journal of Social Psychology, 68,* 1966, 323–330.

Arnqvist, G. Comparative evidence for the evolution of genitalia by sexual selection. *Nature, 393,* 1998, 784–786.

Athanasiou, R. & Greene, P. Physical attractiveness and helping behavior. *Proceedings of the 81st Annual Convention of the American Psychological Association, 8,* 1973, 289–290.

Atzwanger, K. & Grammer, K. Babyness and sexual attraction. Paper available from LBI for Urban Ethology, Vienna, Austria.

Aucoin, K. *The Art of Makeup.* New York: HarperCollins, 1994.

———. *Making Faces.* Boston, MA: Little, Brown, 1997.

Austad, S. Aging and caloric restrictions: Human effects and mode of action. *Neurobiology of Aging, 16,* 1995, 851–852.

Badia, P. N., Wesensten, N., Lammers, W., Culpeper, J. & Harsh, J. Responsiveness to olfactory stimuli presented in sleep. *Physiology and Behavior, 48,* 1990, 87–90.

Baker, M. A. *Sex Differences in Human Performance.* New York: John Wiley, 1987.

Bangs, L. *Psychotic Reactions and Carburetor Dung.* Ed. G. Marcus. New York: Vintage, 1988.

Banner, L. W. *American Beauty.* Chicago, IL: University of Chicago Press, 1983.

Barkow, J. K., Cosmides, L. & Tooby, J., eds. *The Adapted Mind: Evolutionary Psychology and the Generation of Culture.* New York: Oxford University Press, 1992.

Barkowitz, P. & Brigham, J. C. Recognition of faces: Own race bias, incentive and time delay. *Journal of Applied Social Psychology, 12,* 1982, 255–268.

Basolo, A. Female preferences for male sword length in the green swordtail Xiphorus helleri, *Animal Behavior, 40,* 1990, 332–338.

Bassnett, S. *Elizabeth I: A Feminist Perspective.* New York: Berg, 1997.

Bateson, M. C. *With a Daughter's Eye: A Memoir of Margaret Mead and Gregory Bateson.* New York: HarperPerennial, 1994.

Bateson, P. Preferences for cousins in Japanese quail. *Nature, 295,* 1982, 236–237.

———. Preferences for close relations in Japanese quail. In *Acta XIX Congressus Internationalis Ornithologici,* Vol. I. Ed. H. Ouellet. Ottawa: University of Ottawa Press, 1988.

Baudelaire, C. *The Flowers of Evil.* Trans. M. & J. Mathews. New York: New Directions, 1989.

———. *The Painter of Modern Life and Other Essays.* Trans. Jonathan Mayne. New York: Da Capo, 1964.

Bayley, S. *Taste: The Secret Meaning of Things.* New York: Pantheon, 1991.

Becker, A. E. *Body, Self, and Society: The View from Fiji.* Philadelphia: University of Pennsylvania Press, 1995.

Becker, G. S. *A Treatise on the Family.* Cambridge: Harvard University Press, 1981.

Bell, Q. *On Human Finery.* 2nd ed. New York: Schocken Books, 1976.

Beloff, H. *Camera Culture.* New York: Basil Blackwell, 1985.

Benson, P. L., Karabenick, S. A. & Lerner, R. M. Pretty pleases: the effects of physical attractiveness, race, and sex on receiving help. *Journal of Experimental Social Psychology, 12,* 1976, 409–415.

Benton, D. The influence of Androstenol—a putative human pheromone—on mood throughout the menstrual cycle. *Biological Psychology, 15,* 1982, 249–256.

Beren, S. E., Hayden, H. A., Wilfley, D. E. & Grilo, C. M. The influence of sexual orientation on body dissatisfaction in adult men and women. *International Journal of Eating Disorders, 20,* 1996, 135–141.

Berry, D. S. & McArthur, L. Z. Perceiving character in faces: The impact of age-related craniofacial changes on social perception. *Psychological Bulletin, 100,* 1986, 3–18.

Bjorntop, P. The association between obesity, adipose tissue distribution and disease. *Acta Medica Scandanvica.* Suppl. J, *723,* 121–134.

Bornstein, M. H., Ferdinandsen, K. & Gross, C. Perception of symmetry in infancy. *Developmental Psychology, 17,* 1981, 82–86.

Boucher, F. *20,000 Years of Fashion: The History of Costume and Adornment.* New York: Harry N. Abrams, 1987.

Brain, R. *The Decorated Body.* London: Hutchinson, 1979.

Bransford, H. *Welcome to Your Facelift: What to Expect Before, During, and After Cosmetic Surgery.* New York: Doubleday, 1997.

Breiter, H. C., Etcoff, N. L., Whalen, P. J., Kennedy, W. A., Rauch, S. L., Buckner, R. L., Strauss, M. M., Hyman, S. E., Rosen, B. R. Response and habituation of the human amygdala during visual processing of facial expression. *Neuron, 17,* 1996, 1–13.

———, Gollub, R. L., Weiskoff, R. M., Kennedy, D. N., Makris, N., Berke, J. D., Goodman, J. M., Kanter, H. L., Gastfriend, D. R., Riorden, J. P., Mathew, R. T., Rosen, B. R. & Hyman, S. E. Acute effects of cocaine on human brain activity and emotion. *Neuron, 29,* 1997, 591–611.

Brigham, J. C. & Barkowitz, P. Do they look alike? The effect of race, sex, experience and attitudes on the ability to recognize faces. *Journal of Applied Social Psychology 8,* 1978, 306–318.

Brink, P. J. The fattening room among the Annang of Nigeria. *Medical Anthropology, 12,* 1989, 131–143.

Brown, P. J. & Konner, M. An anthropological perspective on obesity. *Annals of the New York Academy of Sciences, 499,* 1987, 29–46.

Brown, R. *Social Psychology.* 2nd ed. New York: Free Press, 1986.

Bruch, H. *The Golden Cage: The Enigma of Anorexia Nervosa.* Cambridge: Harvard University Press, 1978.

Brumberg, J. J. *Fasting Girls: The History of Anorexia.* New York: Plume, 1989.

Brundage, L. E., Berlega, V. J. & Cash, T. F. The effects of physical attractiveness and need for approval on self-disclosure. *Personality and Social Psychology Bulletin, 3,* 1977, 63–66.

Buckley, W. A.; Yesalis, C. W., Friedl, K. E., Anderson, W., Streit, A. & Wright, J. Estimated prevalence of anabolic steroid use among male high school seniors. *Journal of the American Medical Association, 260,* 1988, 3441–3445.

Burson, N. *The "age machine" and composite portraiture.* Cambridge, MA: MIT List Visual Arts Center, 1990.

———, Carling, R. & Kramlich, D. *Composites: Computer Generated Portraits.* New York: William Morrow, 1986.

Burt, D. M. & Perrett, D. I. Perceptual asymmetries in face judgements. *Neuropsychologia, 35,* 1997, 685–693.

Burton, A. M., Bruce V. & Dench, N. What's the difference between men and women? Evidence from facial measurement. *Perception, 22,* 1993, 153–176.

Burton, Sir R. & Arbuthnot, F. F., trans. *The Kama Sutra of Vatsayana.* London: Diamond, 1996.

Bushnell, I. W. R., Sai, F. & Mullin, J. T. Neonatal recognition of the mother's face. *British Journal of Developmental Psychology, 7,* 1989, 3–15.

Buss, D. M. *The Evolution of Desire: Strategies of Human Mating.* New York: Basic Books, 1994.

———. Sex differences in human mate preferences: Evolutionary hypotheses tested in 37 cultures. *Behavioral and Brain Sciences, 12,* 1989, 1–14.

——— & Schmitt, D. P. Sexual strategies theory: An evolutionary perspective on human mating. *Psychological Review, 100,* 1993, 204–232.

Butler, S. R., Suskind, M. R. & Schanberg, S. M. Maternal behavior as regulator of polyamine biosynthesis in brain and heart of the rat pup. *Science, 199,* 1978, 445–447.

Bynum, C. W. "The female body and religious practices in the later Middle Ages." In *Fragments for a History of the Human Body.* Part 1. Ed. Michel Feher with Ramona Nadaff and Nadia Tazi. New York: Zone, 1989.

Calden, G., Lundy, R. M. & Schlafer, R. S. Sex differences in body concepts. *Journal of Consulting Psychology, 23,* 1959, 378.

Calhoun, K. H. "Lip Aesthetics." In *Surgery of the Lip.* Ed. K. H. Calhoun & C. M. Steinberg. New York: Thieme, 1992.

Campbell, L. *Renaissance Portraits: European Portrait-Painting in the 14th, 15th, and 16th Centuries.* New Haven: Yale University Press, 1990.

Capote, T. *Breakfast at Tiffany's.* New York: Vintage, 1993.

Caro, T. M. Human breasts: Unsupported hypotheses. *Human Evolution, 2,* 1987, 271–282.

——— & Sellen, D. W. The reproductive advantage of fat in women. *Ethology and Sociobiology, 11,* 1990, 51–66.

Cash, T. F. Losing hair, losing points? The effects of male pattern baldness on social impression formation. *Journal of Applied Social Psychology, 20,* 1990, 154–167.

——— & Kilcullen, R. N. The aye of the beholder: Susceptibility to sexism and beautyism in the evaluation of managerial applicants. *Journal of Applied Social Psychology, 15,* 1985, 591–605.

       & Soloway, D. Self-disclosure and correlates of physical attractiveness: An exploratory study. *Psychological Reports, 36,* 1975, 579–586.

Castiglione, B. *The Courtier.* New York: Penguin, 1976.

Cavalli-Sforza, L. L. & Cavalli-Sforza, F. *The Great Human Diasporas: The History of Diversity and Evolution.* Reading, MA: Addison Wesley, 1995.

Cavell, S. *The World Viewed: Reflections on the Ontology of Film.* New York: Viking Press, 1971.

Chagnon, N. *Yanomamo: The Last Days of Eden.* San Diego: Harcourt Brace Jovanovich, 1992.

Chaiken, S. Communicator physical attractiveness and persuasion. *Journal of Personality and Social Psychology, 37,* 1979, 1387–1397.

Chatwin, B. *What Am I Doing Here?* New York: Penguin, 1989.

Chiroro, P. & Valentine, T. An investigation of the contact hypothesis of the own-race bias in face recognition. *Quarterly Journal of Experimental Psychology, 48A,* 1995, 879–894.

Clark, K. *The Nude: A Study in Ideal Form.* Princeton: Princeton University Press, 1972.

Clifford, M. M. Physical attractiveness and academic performance. *Child Study Journal, 4,* 1975, 201–209.

       & Walster, E. Research note: The effects of physical attractiveness on teacher expectations. *Sociology of Education, 46,* 1973, 248–258.

Cohen, S. B. *Mindblindness: An Essay on Autism and Theory of Mind.* Cambridge: MIT Press, 1995.

Connolly, B. & Anderson, R. *First contact: New Guinea Highlanders Encounter the Outside World.* New York: Viking Penguin, 1987.

Cooper, W. *Hair: Sex, Society, Symbolism.* London, Aldus, 1971.

Corson, R. *Fashions in Hair: The First Five Thousand Years.* London: Peter Owen, 1991.

      . *Fashions in Makeup: From Ancient to Modern Times.* London: Peter Owen, 1972.

Crawley, T. *The Films of Brigitte Bardot.* New York: Citadel Press, 1975.

Cronin, H. *The Ant and the Peacock.* Cambridge: Cambridge University Press, 1991.

Cronin, V. *Louis XIV.* Boston: Houghton Mifflin, 1965.

Cross, J. F. & Cross, J. Age, sex, race and the perception of beauty. *Developmental Psychology, 5,* 1971, 433–439.

Cunningham, M. R. Measuring the physical in physical attractiveness: Quasi-experiments on the sociobiology of female facial beauty. *Journal of Personality and Social Psychology, 50,* 1986, 925–935.

      , Barbee, A. P., Pike, C. L. What do women want? Facialmetric assessment of multiple motives in the perception of male facial physical attractiveness. *Journal of Personality and Social Psychology, 59,* 1990, 61–76.

      , Roberts, A. R., Barbee, A. P., Cruen, P. B. & Wu, C-H. Their ideas of beauty are, on the whole, the same as ours: Consistency and variability in the cross-cultural perception of female physical attractiveness. *Journal of Personality and Social Psychology, 68,* 1995, 261–279.

Dabbs, J. M., Jr. Testosterone, smiling, and facial appearance. *Journal of Nonverbal Behavior, 21,* 1997, 45–55.

Dabbs, M. & Stokes, N. A. Beauty is power: The use of space on the sidewalk. *Sociometry, 38,* 1975, 551–557.

Dalby, L. *Kimono: Fashioning Culture.* New Haven: Yale University Press, 1993.

Daly, M. & Wilson, M. I. Child maltreatment from a sociobiological perspective, *Journal of Marriage and the Family, 43,* 1980, 277–288.

_____ & Wilson, M. I. Whom are newborn babies said to resemble? *Ethology and Sociobiology*, *3*, 1982, 69–78.

Dannenmaier, W. D. & Thumin, F. J. Authority status as a factor in perceptual distortion of size. *Journal of Social Psychology*, *63*, 1964, 361.

Darwin, C. *Autobiography*. Ed. Sir Francis Darwin. New York: Henry Schuman, 1950.

_____. *The Descent of Man, and Selection in Relation to Sex*. Princeton: Princeton University Press, 1981.

_____. *The Expression of the Emotions in Man and Animals: The Definitive Edition*. Introduction, Afterword, and Commentaries by Paul Ekman. New York: Oxford University Press, 1998.

Davis, C., Brewer, H. & Weinstein, M. A study of appearance anxiety in young men. *Social Behavior and Personality*, *21*, 1993, 63–74.

Davis, K. *Reshaping the Female Body: The Dilemmas of Cosmetic Surgery*. New York: Routledge, 1995.

Dawkins, R. *The Blind Watchmaker*. New York: W. W. Norton, 1987.

_____. *River Out of Eden: A Darwinian View of Life*. New York: Basic Books, 1995.

Deaux, K. & Hanna, R. Courtship in the personals column: The influence of gender and sexual orientation. *Sex Roles*, *11*, 1984, 363–375.

Dermer, M. & Thiel, D. L. When beauty may fail. *Journal of Personality and Social Psychology*, *31*, 1975, 1168–1176.

DeSantis, A. & Kayson, W. A. Defendants' characteristics of attractiveness, race, and sex and sentencing decisions. *Psychological Reports*, *81*, 1997, 679–683.

Deutsch, F. M., LeBaron, D. & Fryer, M. M. What is in a smile? *Psychology of Women Quarterly*, *11*, 1987, 341–352.

_____, Zalenski, C. M. & Clark, M. E. Is there a double standard of aging? *Journal of Applied Social Psychology*, *16*, 1986, 771–785.

Diamond, J. *The Third Chimpanzee: The Evolution and Future of the Human Animal*. New York: HarperCollins, 1992.

_____. *Why Is Sex Fun? The Evolution of Human Sexuality*. New York: Basic Books, 1997.

Diaz, S., Aravena, R., Cordenas, H., Casado, M. & Miranda, P. Contraceptive efficiency of lactational amenorrhea in urban Chilean women. *Contraception*, *43*, 1991, 335–352.

_____, Seron-Ferre, M., Croxatto, H. B. & Veldhuis, J. Neuroendocrine mechanisms of lactational infertility in women. *Biological Research*, *28*, 1995, 155–163.

Dickinson, Emily. *Selected Poems and Letters*. R. N. Linscott, ed. New York: Anchor, 1959.

Diener, E., Wolsic, B. & Fujita, F. Physical attractiveness and subjective well-being. *Journal of Personality and Social Psychology*, *69*, 1995, 120–129.

Dillon, D. J. Measurement of perceived body size. *Perceptual and Motor Skills*, *14*, 1962, 191–196.

Dimberg, U. & Lungquist, L. O. Gender differences in facial reactions to facial expressions. *Biological Psychology*, *30*, 1990, 151–159.

Dion, K. K. Physical attractiveness and evaluation of children's trangressions. *Journal of Personality and Social Psychology*, *24*, 1972, 207–213.

_____ & Berscheid, E. Physical attractiveness and peer perception among children. *Sociometry*, *37*, 1974, 1–12.

_____, Berscheid, E. & Walster, E. What is beautiful is good, *Journal of Personality and Social Psychology*, *24*, 1972, 285–290.

Doty, R. L., Applebaum, L. S., Zusho, H. & Settle, R. G. Sex differences in odor identification ability: a cross-cultural analysis. *Neuropsychologia*, *23*, 1985, 667–672.

————, Snyder, P. J., Huggins, G. R. & Lowry, L. D. Endocrine, cardiovascular, and psychological correlates of olfactory sensitivity changes during the human menstrual cycle. *Journal of Comparative Physiology and Psychology, 95,* 1981, 45–51.

Drewnowski, A., Lee, D. K. & Kahn, D. D. Bulimia in college women: Incidence and recovery rates. *American Journal of Psychiatry, 145,* 1988, 753–755.

Dugatkin, L. A. & Sargent, R. C. Male-male association patterns and female proximity in the guppy Poecilia reticulata. *Behavioral Ecology and Sociobiology, 35,* 1994, 141.

Dunn, L. B., Damesyn, M., Moore, A. A., Reuben, D. B. & Greendale, G. A. Does estrogen prevent skin aging? Results from the first National Health and Nutrition Examination Survey. *Archives of Dermatology, 133,* 1997, 339–342.

DuRant, R. H., Rickert, V. I., Ashworth, C. S., Newman, C. & Slavens, G. Use of multiple drugs among adolescents who use anabolic steroids. *New England Journal of Medicine, 328,* 1993, 922–926.

Dutton, K. R. *The Perfectible Body: The Western Ideal of Physical Development.* London: Cassell, 1995.

Eagly, A. H., Ashmore, R. D., Makhijani, M. G. & Longo, L. C. What is beautiful is good, but . . . : A meta-analytic review of research on the physical attractiveness stereotype. *Psychological Bulletin, 110,* 1991, 109–128.

Eberhard, W. G. *Sexual Selection and Animal Genitalia.* Cambridge: Harvard University Press, 1985.

Eco, U. *Art and Beauty in the Middle Ages.* New Haven: Yale University Press, 1986.

Eibl-Eibesfeldt, I. *Love and Hate.* New York: Holt, Rinehart and Winston, 1971.

Ekman, P. *Darwin and Facial Expression: A Century of Research in Review* (New York: Academic Press, 1973).

————. "Universality of emotional expression? A personal history of the dispute." In C. Darwin, 1998.

————. "Universals and cultural differences in facial expressions of emotion." In *Nebraska Symposium on Motivation.* Ed. J. Cole. Lincoln: University of Nebraska Press, 1971.

———— & Friesen, W. V. *Unmasking the Face.* Palo Alto: Consulting Psychologists Press, 1984.

————, Hager, J. C. & Friesen, W. V. The symmetry of emotional and deliberate facial actions. *Psychophysiology, 18,* 1981, 101–105.

Elder, G. H. Appearance and education in marriage mobility. *American Sociological Review, 34,* 1969, 519–533.

Elias, N. *The Court Society.* Oxford: Basil Blackwell, 1983.

Eliot, G. *Adam Bede.* New York: Penguin, 1981.

Ellis, B. J. & Symons, D. Sex differences in sexual fantasy: An evolutionary psychological approach. *Journal of Sex Research, 27,* 1990, 527–555.

Enlow, D. H. *Handbook of Facial Growth.* Philadelphia: W. B. Saunders, 1982.

Ernster, V. L., Grady, D., Miike, R., Black, D., Selby, J., Kerlikowske, K. Facial wrinkling in men and women by smoking status. *American Journal of Public Health, 85,* 1995, 78–82.

Etcoff, N. L. Asymmetries in recognition of emotion. In *Handbook of Neuropsychology.* Vol. 3. Ed. F. Boller and J. Grafman (Amsterdam: Elsevier, 1989).

————. Beauty and the beholder. *Nature, 368,* 1994, 186–187.

————. Perceptual and conceptual organization of facial emotions: Hemispheric differences. *Brain and Cognition, 3,* 1984, 385–412

————. Selective attention to facial identity and facial emotion. *Neuropsychologia, 22,* 1984, 281–295.

————, Freeman, R. & Cave, K. Can we lose memories of faces? Content specificity and awareness in a prosopagnosic. *Journal of Cognitive Neuroscience, 3,* 1991, 25–41.

Evans, D. J., Barth, J. H. & Burkem, C. W. Body fat topography in women with androgen excess. *International Journal of Obesity, 12,* 1988, 157–162.

Fallon, A. E. & Rozin, P. Sex differences in perception of desirable body shape. *Journal of Abnormal Psychology, 94,* 1985, 102–105.

Farkas, L. G., Hreczko, T. A., Kolar, J. C. & Munro, I. R. "Vertical and horizontal proportions of the face in young adult North American Caucasians: Revisions of neoclassical canons. *Plastic and Reconstructive Surgery, 75,* 1985, 328–338.

————, Munro, I. R. & Kolar, J. C. The validity of neoclassical facial proportion canons. In *Anthropometric Facial Proportions in Medicine.* Ed. L. G. Farkas & I. R. Munro. Springfield, IL: Charles C. Thomas, 1987.

Feingold, A. Good looking people are not what we think. *Psychological Bulletin, 111,* 1992, 304–341.

————. Matching for attractiveness in romantic partners and same-sex friends: A meta-analysis and theoretical critique. *Psychological Bulletin, 104,* 1988, 226–235.

———— & Mazzella, R. Gender differences in body image are increasing. *Psychological Science, 9,* 1998, 190–195.

Feinman, S. & Gill, G. W. Sex differences in physical attractiveness preferences. *Journal of Social Psychology, 105,* 1978, 43–52.

Ferrario, V. F., Sforza, C., Pizzini, G., Vogel, G. & Miani, A. Sexual dimorphism in the human face assessed by euclidean matrix analysis. *Journal of Anatomy, 183,* 1993, 593–600.

Findlay, G. H. & de Beer, H. A. Chronic hydroquinone poisoning of the skin from skin-lightening cosmetics: A South African epidemic of ochronosis of the face in dark-skinned individuals. *South African Medical Journal, 57,* 1980, 187–190.

Fisher, H. *Anatomy of Love: The Mysteries of Mating, Marriage, and Why We Stray.* New York: Fawcett Columbine, 1992.

Fisher, R. A. *The Genetical Theory of Natural Selection.* 2nd ed. New York: Dover, 1958.

Fisher, W. A., Branscombe, N. R. & Lemery, C. R. The bigger the better? Arousal and attributional responses to erotic stimuli that depict different penis sizes. *Journal of Sex Research, 19,* 1983, 377–396.

Fleming, I. *Goldfinger.* New York: Macmillan, 1959.

Flugel, J. C. *The Psychology of Clothes.* London: Hogarth Press, 1966.

Fombonne, E. Anorexia nervosa: No evidence of an increase. *British Journal of Psychiatry, 166,* 1995, 462–471.

Ford, C. S. & Beach, F. A. *Patterns of Sexual Behavior.* New York: Harper and Row, 1951.

Ford, E. S., Freedman, D. S. & Byers, T. Baldness and ischemic heart disease in a national sample of men. *American Journal of Epidemiology, 143,* 1996, 651–657.

France, A. *Penguin Island.* Trans. A. W. Evans. Norwalk, CT: Heritage Press, 1975.

Frank, R. & Cook, P. J. *The Winner-Take-All Society.* New York: Penguin, 1995.

Fraser, K. *Scenes from the Fashionable World.* New York: Alfred Knopf, 1987.

————. Introduction. In Liberman and Wintour, 1992.

Frayser, S. *Varieties of Sexual Experience: An Anthropological Perspective on Human Sexuality.* New Haven: HRAF Press, 1985.

Freud, S. *Civilization and Its Discontents.* Trans. James Strachey. New York: W. W. Norton, 1961.

————. *Three Essays on the Theory of Sexuality.* Trans. James Strachey. New York: Basic Books, 1962.

Freund, R. M. & Nolan, W. B. Correlation between brow lift outcomes and aesthetic ideals for eyebrow height and shape in females. *Plastic and Reconstructive Surgery, 97*, 1996, 1343–1348.

Frieze, I. H., Olson, J. E. & Good, D. C. Perceived and actual discrimination in the salaries of male and female managers. *Journal of Applied Social Psychology, 20*, 1990, 46–67.

Frisch, R. E. Fatness and fertility. *Scientific American, 258*, 1988, 70–77.

Frost, P. Human skin color: A possible relationship between its sexual dimorphism and its social perception. *Perspectives in Biology and Medicine, 32*, 1988, 38–58.

_____. Human skin color: The sexual differentiation of its social perception. *Mankind Quarterly, 30*, 1989, 3–16.

Furnham, A. & Alibhai, N. Cross-cultural differences in the perception of female body shapes. *Psychological Medicine, 13*, 1983, 829–837.

_____ & Baguma, P. Cross-cultural differences in the evaluation of male and female body shapes. *International Journal of Eating Disorders, 15*, 1994, 81–89.

Fussell, S. W. *Muscle: The Confessions of an Unlikely Bodybuilder.* New York: Poseidon Press, 1991.

Galton, F. Composite portraits. *Nature, 18*, 1878, 97–100.

_____. Generic Images. *Proceedings of the Royal Institution, 9*, 1879, 161–170.

_____. *Inquiries into Human Faculty and Its Development.* London: Macmillan, 1883.

Gangestad, S. W. & Buss, D. M. Pathogen prevalence and human mate preferences. *Ethology and Sociobiology, 14*, 1993, 89–96.

_____, Thornhill, R. & Yeo, R. A. Facial attractiveness, developmental stability, and fluctuating asymmetry. *Ethology and Sociobiology, 15*, 1994, 73–85.

Gardner, M. The cult of the golden ratio. *Skeptical Inquirer, 18*, 1994, 243–247.

Garn, S. M. From the Miocene to olestra: a historical perspective on fat consumption. *Journal of the American Dietary Association, 97*, 1997, S54–57.

_____, Sullivan, T. V. & Hawthorne, V. M. Education level, fatness, and fatness differences between husbands and wives. *American Journal of Clinical Nutrition, 50*, 1989, 749–745.

Gedo, J. E. *Portraits of the Artists: Psychoanalysis of Creativity and Its Vicissitudes.* New York: Guilford, 1983.

Ghyka, M. *The Geometry of Art and Life.* New York: Dover, 1977.

Giddon, D. B. Orthodontic applications of psychological and perceptual studies of facial esthetics. *Seminars in Orthodontics, 1*, 1995, 82–93.

_____, Bernier, D. L., Evans, C. A. & Kinchen, J. A. Comparison of two computer-animated imaging programs for quantifying facial profile preference. *Perceptual and Motor Skills, 82*, 1996, 1251–1264.

Gilbert, C. & Bakan, P. Visual asymmetry in perception of faces. *Neuropsychologia, 11*, 1973, 355–362.

Gillis, J. S. *Too tall, too small.* Champaign, IL: Institute for Personality and Ability Testing, 1982.

_____ & Avis, W. E. The male-taller norm in mate selection. *Personality and Social Psychology Bulletin, 6*, 1980, 396–401.

Goffman, E. Symbols of class status. *British Journal of Sociology, 2*, 1951, 294–304.

Goin, J. M. & Goin, M. C. *Changing the Body: Psychological Effects of Plastic Surgery.* Baltimore: Williams and Wilkins, 1981.

Goldstein, A. G. & Papageorge, J. Judgments of facial attractiveness in the absence of eye movements. *Bulletin of the Psychonomic Society, 15*, 1980, 269–70.

Gombrich, E. H. The mask and the face: The perception of physiognomic likeness in life and in art. In *Art, Perception, and Reality*. Ed. E. H. Gombrich, J. Hochberg & M. Black. Baltimore: Johns Hopkins University Press, 1972.

Goodall, J. V. L. The behavior of free-living chimpanzees in the Gombe stream area. *Animal Behavior Monographs, 1*, 1968, 161–311.

Goodman, A. H. "Bred in the bone? *The Sciences*, March/April 1997, 20–25.

Gordon, R. A. *Anorexia and Bulimia: Anatomy of a Social Epidemic*. Cambridge: Basil Blackwell, 1990.

Goren, C. C., Sarty, M. & Wu, P. Y. K. Visual following and pattern discrimination of face-like stimuli by newborn infants. *Pediatrics, 56*, 1975, 544–549.

Gould, S. J. "A biological homage to Mickey Mouse." In S. J. Gould, *The Panda's Thumb: More Reflections in Natural History*. New York: W. W. Norton, 1980.

———. "Petrus Camper's angle." In S. J. Gould, *Bully for Brontosaurus: Reflections in Natural History*. New York: W. W. Norton, 1992.

Graddol, D. & Swann, J. *Gender Voices*. Oxford: Basil Blackwell, 1989.

Grady, D. & Ernster, V. L. Does cigarette smoking make you old and ugly? *American Journal of Epidemiology, 135*, 1995, 839–842.

Grammer, K. "5-a-androst-162n-3a-on: A male pheromone? A brief report." *Ethology and Sociobiology, 14*, 1993, 201–207.

——— & Thornhill, R. Human (Homo sapiens) facial attractiveness and sexual selection: The role of symmetry and averageness. *Journal of Comparative Psychology, 108*, 1994, 233–242.

Graziano, W. G., Brothen, T. & Berscheid, E. Height and attraction: Do men see women eye to eye? *Journal of Personality, 46*, 1978, 128–145.

———, Jensen-Campbell, L. A., Shebilske, L. J. & Lundgren, S. R. Social influence, sex differences, and judgments of beauty: Putting the interpersonal back in interpersonal attraction. *Journal of Personality and Social Psychology, 65*, 1993, 522–531.

Green, C. D. All that glitters: A review of psychological research on the aesthetics of the golden section. *Perception, 24*, 1995, 937–968.

Gross, M. *Model: The Ugly Business of Beautiful Women*. New York: William Morrow, 1995.

Gutek, B. A. *Sex and the Workplace: The Impact of Sexual Behavior and Harassment on Women, Men, and the Organization*. San Francisco: Jossey-Bass, 1985.

Gwynne, D. T. Genitally does it. *Nature, 393*, 1998, 734–735.

Haight, G. *George Eliot: A Biography*. London: Penguin, 1968.

———. *George Eliot Letters*. London: Oxford University Press, 1954.

Haiken, E. *Venus Envy: A History of Cosmetic Surgery*. Baltimore: Johns Hopkins University Press, 1997.

Haith, M. M., Bergman, T. & Moore, M. J. Eye contact and face scanning in early infancy. *Science, 198*, 1977, 853–855.

Hamermesh, D. S. & Biddle, J. E. Beauty and the labor market. *American Economic Review, 84*, 1994, 1174–1194.

Hamilton, W. D. & Zuk, M. Heritable true fitness and bright birds: A role for parasites. *Science, 218*, 1982, 384–387.

Harris, M. *Our Kind: Who We Are, Where We Came From, and Where We Are Going*. New York: HarperPerennial, 1989.

Hart, K. J. & Ollendick, T. H. Prevalence of bulimia in working and university women. *American Journal of Psychiatry, 142*, 1985, 851–853.

Hartnett, J. J., Bailey, K. G. & Hartley, C. Body height, position, and sex as determinants of personal space. *Journal of Psychology, 87*, 1974, 129–136.

Hasky, J. Social class differentials in remarriage after divorce: Results from a forward linkage study. *Population Trends, 47*, 1987, 34–42.

Hatala, M. N. & Prehodka, J. Content analysis of gay male and lesbian personal advertisements. *Psychological Reports, 78,* 1996, 371–374.

Hatfield, E. & Sprecher, S. *Mirror, Mirror: The Importance of Looks in Everyday Life.* Albany: State University of New York Press, 1986.

Heilman, M. E. & Saruwatari, L. R. When beauty is beastly: The effects of appearance and sex on evaluations of job applicants for managerial and nonmanagerial jobs. *Organizational Behavior and Human Performance, 23,* 1979, 360–372.

Held, B. L., Nader, S., Rodriguez-Rigau, L. J., Smith, K. D. & Steinberger, E. Acne and hyperandrogenism. *Journal of the American Academy of Dermatology, 10,* 1984, 223–226

Hensley, W. E. & Cooper, R. Height and occupational success: A review and critique. *Psychological Reports, 60,* 1987, 843–849.

Hersey, G. *The Evolution of Allure: Sexual Selection from the Medici Venus to the Incredible Hulk.* Cambridge: MIT Press, 1996.

Hess, E. H. Attitude and pupil size. *Scientific American, 212,* 1965, 46–54.

_____ & Polt, J. H. Pupil size related to interest value of visual stimuli. *Science, 132,* 1960, 349–350.

Hess-Biber, S. *Am I Thin Enough Yet? The Cult of Thinness and the Commercialization of Identity.* New York: Oxford University Press, 1996.

Hickey, D. *The Invisible Dragon: Four Essays on Beauty.* Los Angeles: Art Issues Press, 1994.

Hinsz, V. B. Facial resemblance in engaged and married couples. *Journal of Social and Personal Relationships, 6,* 1989, 223–229.

Hirukawa, T. & Yamaguchi, M. Effect of sexual dimorphism on human facial attractiveness. *ATR HIP Research Laboratories,* 1996.

Hoare, P. *Serious Pleasures: The Life of Stephen Tennant.* New York: Hamish Hamilton, 1991.

Hoetink, H. *The Two Variants in Caribbean Race Relations: A Contribution to the Sociology of Segmented Societies.* New York: Oxford University Press, 1967.

Hofstadter, A. & Kuhns, R., eds. *Philosophies of Art and Beauty: Selected Readings in Aesthetics from Plato to Heidegger.* Chicago: University of Chicago Press, 1964.

Hoglund, J., Eriksson, M. & Lindell, L. E. Females of the lek-breeding great snipe, Gallinago media, prefer white tails. *Animal Behavior, 40,* 1990, 23–32.

Hollander, A. *Seeing Through Clothes:* Berkeley: University of California Press, 1993.

Horai, J., Naccari, N., Fatoullen, E. The effects of expertise and physical attractiveness upon opinion agreement and liking. *Sociometry, 37,* 1974, 601–606.

Horvath, T. Correlates of physical beauty in men and women. *Social Behavior and Personality, 7,* 1979, 145–151.

Huntley, H. E. *The Divine Proportion: A Study in Mathematical Beauty.* New York: Dover, 1970.

Iliffe, A. H. A study of preferences in feminine beauty. *British Journal of Psychology, 51,* 1960, 267–273.

Jackson, D. J. & Huston, T. L. Physical attractiveness and assertiveness. *Journal of Social Psychology, 96,* 1975, 79–84.

Jackson, L. A. *Physical Appearance and Gender: Sociobiological and Sociocultural Perspectives.* Albany: State University of New York Press, 1992.

_____ & Cash, T. F. Components of gender stereotypes and their implications for inferences on stereotypic and nonstereotypic dimensions. *Personality and Social Psychology Bulletin, 11,* 1985, 326–344.

_____ & Ervin, K. S. Height stereotypes of women and men: The liability of shortness for both sexes. *Journal of Social Psychology, 132,* 1991, 433–445.

———, Hunter, J. E. & Hodge, C. N. Physical attractiveness and intellectual competence: A meta-analytic review. *Social Psychology Quarterly, 58,* 1995, 108–122.

Jacobson, E. *The Self and the Object World.* New York: International Universities Press, 1964.

Jacobson, W. E., Edgerton, M. T., Meyer, E., Canter, A. & Slaughter, R. "Psychiatric evaluation of male patients seeking cosmetic surgery." *Plastic and Reconstructive Surgery, 26,* 1960, 356–371.

Jaffe, B. & Fanshel, D. *How They Fared in Adoption: A Follow-up Study.* New York: Columbia University Press, 1970.

James, H. *Selected Letters.* Ed. Leon Edel. Cambridge: Harvard University Press, 1987.

Jankowiak, W. R., Hill, E. M. & Donovan, J. M. The effects of sex and sexual orientation on attractiveness judgments: An evolutionary interpretation. *Ethology and Sociobiology, 12,* 1992, 73–85.

Johnston, V. S. & Franklin, M. Is beauty in the eye of the beholder? *Ethology and Sociobiology, 14,* 1993, 183–199.

———, & Oliver-Rodriguez, J. C. Facial beauty and the late positive component of event-related potentials. *Journal of Sex Research, 34,* 1997, 188–198.

Jones, D. M. *The Evolutionary Psychology of Human Physical Attractiveness: Results from Five Populations.* Ph.D. Diss. University of Michigan, Ann Arbor, 1994.

———. Sexual selection, physical attractiveness, and facial neoteny. *Current Anthropology, 36,* 1995, 723–748.

——— & Hill, K. Criteria of facial attractiveness in five populations. *Human Nature, 4,* 1993, 271–296.

Joyce, J. *A Portrait of the Artist as a Young Man.* New York: Viking Press, 1971.

Kant, I. *Observations on the Feeling of the Beautiful and the Sublime.* Trans. J. T. Goldthwait. Berkeley: University of California Press, 1960.

Kaplan, R. M. Is beauty talent? Sex interaction in the attractiveness halo effect. *Sex Roles, 4,* 1978, 195–204.

Keating, C., Mazur, A., Segall, M. A cross-cultural exploration of physiognomic traits of dominance and happiness. *Ethology and Sociobiology, 2,* 1981, 41–48.

Keil, F. *Concepts, Kinds, and Conceptual Development.* Cambridge: MIT Press, 1989.

Kelsh, N. *Naked Babies.* New York: Penguin Studio, 1996.

Kenrick, D. T. & Gutierres, S. E Contrast effects and judgments of physical attractiveness: When beauty becomes a social problem. *Journal of Personality and Social Psychology, 38,* 1980, 131–140.

———, Gutierres, S. E. & Goldberg, L. L. Influences of popular erotica on judgments of strangers and mates. *Journal of Experimental Social Psychology, 25,* 1989, 159–167.

Keyes, R. *The Height of Your Life.* Toronto: Little Brown, 1980.

Kirkpatrick, M. Sexual selection: Is bigger always better? *Nature, 337,* 1989, 116.

Kirk-Smith, M. D. & Booth, D. A. "Effects of Androstenone on choice of location in others' presence." In *Olfaction and Taste.* Vol. 7. Ed. H. van der Starre. London: IRL Press, 1980.

———, Booth, D. A., Carroll, D. & Davies, P. Human social attitudes affected by androstenol. *Research Communications in Psychology, Psychiatry, and Behavior, 3,* 1978, 379–384.

Knecht, H., Pike-Tay, A., White, R. *Before Lascaux: The Complex Record of the Early Upper Paleolithic.* Boca Raton: CRC Press, 1993.

Knight, C., Powers, C. & Watts, I. The human symbolic revolution; A Darwinian account. *Cambridge Archaeological Journal, 5,* 1995, 75–114.

Koch, R. J., Troell, R. J., Goode, R. L. Contemporary management of the aging brow and forehead. *Laryngoscope, 107*, 1997, 710–715.

Koerner, J. L. *The Moment of Self-Portraiture in German Renaissance Art.* Chicago: University of Chicago Press, 1993.

Koeslag, J. H. Koinophilia groups sexual creatures into species, promotes stasis, and stabilizes social behavior. *Journal of Theoretical Biology, 144*, 1990, 15–35.

Komori, T., Fujiwara, R., Tanida, M., Nomura, J. & Yokoyama, M. M. Effects of citrus fragance on immune function and depressive status. *Neuroimmunomodulation, 2*, 1995, 174–180.

Konner, M. *The Tangled Wing: Biological Constraints on the Human Spirit.* New York: Harper Colophon, 1982.

Kowner, R. Facial asymmetry and attractiveness judgments in developmental perspective. *Journal of Experimental Psychology: Human Perception and Performance, 22*, 1996, 662–675.

Kramer, P. D. *Listening to Prozac: A Psychiatrist Explores Antidepressant Drugs and the Remaking of the Self.* New York: Penguin, 1993.

Krebs, D. & Adinolfi, A. A. Physical attractiveness, social relations, and personality style. *Journal of Personality and Social Psychology, 31*, 1975, 245–253.

Kundera, M. *The Book of Laughter and Forgetting.* New York: Penguin, 1981.

Kurtz, D. L. Physical appearance and stature: Important variables in sales recruiting. *Personnel Journal, 48*, 1969, 981–983.

Lakoff, R. T. & Scherr, R. L. *Face Value: the Politics of Beauty.* Boston: Routledge & Kegan Paul, 1984.

Lambert, E. Z. *The Face of Love: Feminism and the Beauty Question.* Boston: Beacon Press, 1995.

Landy, D. & Sigall, H. Beauty is talent: Task evaluation as a function of the performer's physical attractiveness, *Journal of Personality and Social Psychology, 29*, 1974, 299–304.

Langlois, J. H., Ritter, J. M., Casey, R. J. & Sawin, D. B. Infant attractiveness predicts maternal behaviors and attitudes. *Developmental Psychology, 31*, 1995, 464–472.

———, Ritter, J. M., Roggman, L. A. & Vaughn, L. S. Facial diversity and infant preferences for attractive faces. *Developmental Psychology, 27*, 1991, 79–84.

——— & Roggman, L. A. Attractive faces are only average. *Psychological Science, 1*, 1990, 115–121.

———, Roggman, L. A. Casey, R. J., Ritter, J. M., Rieser-Danner, L. A. & Jenkins, V. Y. Infant preferences for attractive faces: Rudiment of a stereotype? *Developmental Psychology, 23*, 1987, 363–369.

———, Roggman, L. A. & Musselman, L. What is average and what is not average about attractive faces. *Psychological Science, 5*, 1994, 214–220.

———, Roggman, L. A. & Rieser-Danner, L. A. Infants' differential social response to attractive and unattractive faces. *Developmental Psychology, 26*, 1990, 153–159.

Lappe, M. *The Body's Edge: Our Cultural Obsession with Skin.* New York: Henry Holt, 1996.

Lauder, E. *Estee: A Success Story.* New York: Random House, 1985.

Laver, J. *Costume and Fashion: A Concise History.* New York: Thames & Hudson, 1985.

Lavrakas, P. J. Female preference for male physiques. *Journal of Research in Personality, 9*, 1975, 324–334.

Lehrman, K. *The Lipstick Proviso: Women, Sex, and Power in the Real World.* New York: Anchor, 1997.

Lesko, S. M., Rosenberg, L. & Shapiro, S. A case-control study of baldness in relation to myocardial infarction in men. *Journal of the American Medical Association, 269,* 1993, 998–1003.

Lewontin, R. *Human Diversity.* New York: Scientific American Library, 1995.

Liberman, A. & Wintour, A., eds. *On the Edge: Images from 100 Years of Vogue.* New York: Random House, 1992.

Lindzey, G. "Morphology and Behavior." In *Theories of Personality: Primary Sources and Research.* 2nd ed. Ed. G. Lindzey, C. S. Hall & M. Manosevitz. New York: John Wiley, 1973.

Lipovetsky, G. *The Empire of Fashion: Dressing Modern Democracy.* Trans. Catherine Porter. Princeton: Princeton University Press, 1994.

Liu, D., Diorio, J., Tannenbaum, B., Cadji, C., Francis, D., Freedman, A., Sharma, S., Pearson, D., Plotsky, P. M. & Meaney, J. M. Maternal care, hippocampal glucocorticoid receptors, and hypothalamic-pituitary-adrenal response to stress. *Science, 277,* 1997, 1659–1662.

Loh, E. S. The economic effects of physical appearance. *Social Science Quarterly, 74,* 1993, 420–438.

Lord, M. G. *Forever Barbie: The Unauthorized Biography of a Real Doll.* New York: William Morrow, 1994.

Low, B. S., Alexander, R. D. & Noonan, K. M. Human hips, breasts and buttocks: Is fat deceptive? *Ethology and Sociobiology, 8,* 1986, 249–257.

Lowe, N. J., Maxwell, A. & Harper, H. Botulinum A exotoxin for glabellar folds: A double-blind, placebo-controlled study with an electromyographic injection technique. *Journal of the American Academy of Dermatology, 35,* 1996, 569–572.

Lucker, W., Beane, W. E. & Hemreich, R. L. The strength of the halo effect in physical attractiveness research. *Journal of Psychology, 107,* 1981, 69–75.

Lurie, A. *The Language of Clothes.* New York: Random House, 1981.

Lykken, D. & Tellegen, A. Happiness is a stochastic phenomenon. *Psychological Science, 7,* 1996, 186–189.

Lynn, M. & Shurgot, B. A. Responses to lonely hearts advertisements: effects of reported physical attractiveness, physique and coloration. *Personality and Social Psychology Bulletin, 10,* 1984, 349–357.

Lyon, B. E., Eadier, J. M. & Hamilton, C. D. Parental choice selects for ornamental plumage in American coot chicks. *Nature, 371,* 1993, 240–243.

Mackenzie, M. The pursuit of slenderness and addiction to self-control. In *Nutrition Update.* Vol. 2. Ed. J. Weininger and G. M. Briggs. New York: John Wiley and Sons, 1985.

Madsen. A. *Chanel: A Woman of Her Own.* New York: Henry Holt, 1990.

Magli, P. "The face and the soul." In *Fragments for a History of the Human Body.* Part 2. Ed. Michel Feher with Ramona Naddaff and Nadia Tazi. New York: Zone, 1989.

Maheux, R., Naud, F., Rioux, M., Grenier, R., Lemay, A. & Langevin, M. A randomized, double-blind, placebo-controlled study on the effect of conjugated estrogens on skin thickness. *American Journal of Obstetrics and Gynecology, 170,* 1994, 642–649.

Maier, R. A., Holmes, D. L., Slaymaker, F. L. & Reich, J. N. The perceived attractiveness of preterm infants. *Infant Behavior and Development, 7,* 1984, 403–414.

———— & Lavrakas, P. J. Attitudes toward women, personality rigidity, and idealized physique preferences in males. *Sex Roles, 11,* 1984, 425–433.

Malinowski, B. *The Sexual Life of Savages in North-Western Melanesia.* New York: Harcourt, Brace, and World, 1929.

Mann, J. Nurturance or negligence: Maternal psychology and behavioral preference among preterm infants. In Barkow, Cosmides and Tooby, 1992.

Mann, S. "Smart Clothing: The Wearable Computer and Wearcam," *Personal Technologies*, 1, March 1997.

Mann, T. *Death in Venice*. New York: Bantam, 1971.

Manning, J. T. Fluctuating asymmetry and body weight in men and women: Implications for sexual selection. *Ethology and Sociobiology*, *15*, 1995, 145–153.

Marks, N. F. Flying solo at midlife: Gender, marital status, and psychological well-being. *Journal of Marriage and the Family*, *58*, 1996, 917–932.

Marks, D. V. P. *Human Beauty: An Economic Analysis*. Ph.D. Thesis, Harvard University, Cambridge, MA, 1989.

Markusson, E. & Folstad, I. Reindeer antlers: Visual indicators of individual quality? *Oecologia*, *110*, 1997, 501–507.

Martel, L. F. & Biller, H. B. *Stature and Stigma: The Biopsychosocial Development of Short Males*. Lexington, MA: Lexington Books, 1987.

Masters, W. H. & Johnson, V. E. *Human Sexual Response*. Boston: Little Brown, 1966.

Matory, W. E., Jr. "Definitions of Beauty in the Ethnic Patient." In *Ethnic Considerations in Facial Aesthetic Surgery*. Ed. W. E. Matory, Jr. Philadelphia: Lippincott-Raven, 1998.

Maurer, D. & Young, R. Newborns' following of natural and distorted arrangements of facial features. *Infant Behavior and Development*, *6*, 1983, 127–131.

Mazella, R. & Feingold, A. The effects of physical attractiveness, race, socioeconomic status, and gender of defendants and victims on judgments of mock jurors: a meta-analysis. *Journal of Applied Social Psychology*, *24*, 1994, 1315–1344.

Mazur, A. U.S. Trends in feminine beauty and overadaptation. *Journal of Sex Research*, *22*, 1986, 281–303.

———, Halpern, C., Udry, J. R. Dominant looking male teenagers copulate earlier. *Ethology and Sociobiology*, *15*, 1994, 87–94.

———, Mazur, J., Keating, C. Military rank attainment of a West Point class: Effects of cadets' physical features. *American Journal of Sociology*, *90*, 1984, 125–150.

McCabe, V. "Facial proportions, perceived age, and caregiving." In *Social and Applied Aspects of Perceiving Faces*. Ed. T. R. Alley. Hillsdale, NJ: Erlbaum, 1988.

McCracken, G. *Big Hair: A Journey into the Transformation of Self*. Woodstock, NY: Overlook Press, 1995.

McGinnis, J. *The Selling of the President*. New York: Andre Deutsch, 1976.

McGuire, M. T. *Darwinian Psychiatry*. New York: Oxford University Press, 1998.

McManus, I. C. & Humphrey, N. K. Turning the left cheek. *Nature*, *243*, 1973, 271–272.

Meerdink, J. E., Garbin, C. P. & Leger, D. W. Cross-gender perceptions of facial attributes and their relation to attractiveness: Do we see them differently than they see us? *Perception and Psychophysics*, *48*, 1990, 227–233.

Meltzoff, A. N. & Moore, M. K. Imitation of facial and manual gestures by human neonates. *Science*, *198*, 1977, 75–78.

Menkin, J., Trussell, J. & Larsen, U. Age and infertility. *Science*, *233*, 1986, 1389–1394.

Michael, R. T., Gagnon, J. H., Laumann, E. O. & Kolata, G. *Sex in America: A Definitive Survey*. Boston: Little Brown, 1994.

Mifflin, M. *Bodies of Subversion: A Secret History of Women and Tattoo*. New York: Juno Books, 1997.

Miller, T. *How to Want What You Have: Discovering the Magic and Grandeur of Ordinary Existence*. New York: Avon, 1995.

Miller, W. I. *The Anatomy of Disgust*. Cambridge: Harvard University Press, 1997.

Minsky, M. Negative expertise. *International Journal of Expert Systems*, *7*, 1994, 13–19.

Mintz, L. B. & Betz, N. E. Sex differences in the nature, realism, and correlates of body image. *Sex Roles, 15*, 1986, 185–195.

Mitchell, M. *Gone With the Wind.* New York: MacMillan, 1938.

Mithen, S. *The Prehistory of the Mind: The Cognitive Origins of Art, Religion, and Science.* London: Thames and Hudson, 1996.

Moghaddam, B. & Pentland, A. Probabilistic visual learning for object representation. *IEEE Transactions on Pattern Analysis and Machine Intelligence, 7*, July 1977, 696–710.

Moller, A. P. Bumblebee preference for symmetrical flowers. *Proceedings of the National Academy of Science, 92*, 1995, 2288–2292.

———. Female choice selects for male sexual tail ornaments in the monogamous swallow. *Nature, 332*, 1988, 640–642.

———. Female swallow preference for symmetrical male sexual ornaments. *Nature, 357*, 1992, 238–240.

———. Male ornament size as a reliable cue to enhanced offspring viability in the barn swallow. *Proceedings of the National Academy of Science, 91*, 1994, 6929–6932.

——— & Eriksson, M. Patterns of fluctuating asymmetry in flowers: Implications for sexual selection in plants. *Journal of Evolutionary Biology, 7*, 1994, 97–113.

———, Soler, M. & Thornhill, R. Breast asymmetry, sexual selection, and human reproductive success. *Ethology and Sociobiology, 16*, 1995, 207–219.

——— & Thornhill, R. Bilateral symmetry and sexual selection: A meta-analysis. *American Naturalist, 151*, 1998, 174–192.

Molloy, J. T. *Dress for Success.* New York: Warner, 1975.

———. *The Woman's Dress for Success Book.* New York: Warner, 1978.

Montagna, W. *The Structure and Function of Skin.* New York: Academic Press, 1962.

——— & Ellis, R., eds. *The Biology of Hair Growth.* New York: Academic Press, 1958.

Montague, A. *Touching: The Human Significance of the Skin.* New York: Columbia University Press, 1971.

Montaigne, M. "On Physiognomy." In M. de Montaigne, *Essays.* Trans. J. M. Cohen. New York: Penguin, 1958.

Monti-Bloch, L., Jennings-White, C., Dolberg, D. S. & Berliner, D. L. The human vomeronasal system. *Psychoneuroendocrinology, 19*, 1994, 673–686.

Moore, M. M. Nonverbal courtship patterns in women: Context and consequences. *Ethology and Sociobiology, 6*, 1985, 237–247.

——— & Butler, D. L. Predictive aspects of nonverbal courtship behavior in women. *Semiotica, 76*, 1989, 205–215.

Morris, D. *The Human Zoo.* New York: Dell, 1969.

———. *The Naked Ape: A Zoologist's Study of the Human Animal.* New York: McGraw-Hill, 1967.

Morselli, P. G. The Minotaur syndrome: Plastic surgery of the facial skeleton. *Aesthetic Plastic Surgery, 17*, 1993, 99–102.

Morton, J. & Johnson, M. H. CONSPEC and CONLEARN: A two-process theory of infant face recognition. *Psychological Review, 98*, 1991, 164–181.

Mueller, U. & Mazur, A. Facial dominance of West Point Cadets as a predictor of later military rank. *Social Forces, 74*, 1996, 823–850.

Mueser, K. T., Grau, B. W., Sussman, S. & Rosen, A. J. You're only as pretty as you feel: facial expression as a determinant of physical attractiveness. *Journal of Personality and Social Psychology, 46*, 1984, 469–478.

Murphy, M. J. & Hellkamp, D. T. Attractiveness and personality warmth: Evaluations of paintings rated by college men and women. *Perceptual and Motor Skills, 43*, 1976, 1163–1166.

Murstein, B. I. Physical attractiveness and marital choice. *Journal of Personality and Social Psychology, 22,* 1972, 8–12.

Muscarella, F. & Cunningham, M. R. The evolutionary significance and social perception of male pattern baldness and facial hair. *Ethology and Sociobiology, 17,* 1996, 99–117.

Myers, D. G. & Diener, E. Who is happy? *Psychological Science, 6,* 1995, 10–19.

Nadler, A., Shapira, R. & Ben-Itzhak, S. Good looks may help: Effects of helper's physical attractiveness and sex of helper on males' and females' help-seeking behavior. *Journal of Personality and Social Psychology, 42,* 1982, 90–99.

Nelson, J. F., Karelus, K., Bergman, M. D. & Felicio, L. S. Neuroendocrine involvement in aging: Evidence from studies of reproductive aging and caloric restriction. *Neurobiology of Aging, 16,* 1995, 837–843.

Nesse, R. M. & Williams, G. C. *Why We Get Sick: The New Science of Darwinian Medicine.* New York: Random House, 1994.

O'Toole, A. J., Peterson, J. & Deffenbacher, K. A. An other-race effect for categorising faces by sex. *Perception, 25,* 1996, 669–676.

Ovid. *The Art of Love.* Trans. Rolfe Humphries. Bloomington: Indiana University Press, 1957.

———. *The Erotic Poems.* Trans. Peter Green. New York: Penguin, 1982.

Orians, G. H. & Heerwagen. "Evolved Responses to Landscapes." In Barkow, Cosmides and Tooby, 1992.

Pacteau, F. *The Symptom of Beauty.* Cambridge: Harvard University Press, 1994.

Pagel, M. Parents prefer plumage. *Nature, 371,* 1994, 200.

Paglia, C. *Sex, Art, and American Culture.* New York: Vintage, 1992.

———. *Sexual Personae: Art and Decadence from Nefertiti to Emily Dickinson.* New York: Vintage, 1991.

Panofsky, E. *Meaning in the Visual Arts.* New York: Harmondsworth, 1970.

Parkinson, D., ed. *The Graham Greene Film Reader: Reviews, Essays, Interviews and Film Stories.* New York: Applause, 1995.

Payne, B., Winakor, G., Farrell-Beck, J. *The History of Costume.* 2nd ed. New York: HarperCollins, 1992.

Pentland, A. P. "Smart Rooms, Smart Clothes," *Scientific American,* April 1998, p. 124.

Perrett, D. I., Lee, K. J., Penton-Voak, I., Rowland, D., Yoshikawa, S., Burt, D. M., Henzi, S. P., Castles, D. & Akamatsu, S. Effects of sexual dimorphism on facial attractiveness. *Nature,* in press.

———, May, K. A., Yoshikawa, S. Facial shape and judgements of female attractiveness. *Nature, 368,* 1994, 239–242.

Perrot, P. *Fashioning the Bourgeoisie: A History of Clothing in the Nineteenth Century.* Trans. R. Bienvenu. Princeton: Princeton University Press, 1994.

Petrie, M. Improved growth and survival of offspring of peacocks with more elaborate trains. *Nature, 371,* 1994, 598–599.

———, Halliday, T. R. & Sanders, C. Peahens prefer peacocks with elaborate trains. *Animal Behavior, 41,* 1991, 323–331.

Phillips, K. A. *The Broken Mirror: Understanding and Treating Body Dysmorphic Disorder.* New York: Oxford University Press, 1996.

Pinker, S. *The Language Instinct.* New York: HarperCollins, 1994.

Piper, D. *The English Face.* London: Thames and Hudson, 1957.

Pipher, M. *Reviving Ophelia: Saving the Selves of Adolescent Girls.* New York: Ballantine, 1994.

Plath, S. *The Bell Jar.* New York: Alfred Knopf, 1963.

Polhemus, T. *Style Surfing: What to Wear in the Third Millennium.* London: Thames and Hudson, 1996.

Pope, H., Gruber, A. J., Olivardia, R. & Phillips, K. A. Muscle dysmorphia: An underrecognized form of body dysmorphic disorder. *Psychosomatics, 38,* 1997, 548–557.

———, Katz, D. L. & Hudson, J. I. Anorexia nervosa and "reverse anorexia" among 108 male bodybuilders. *Comprehensive Psychiatry, 34,* 1993, 406–409.

Pound, E. *Gaudier-Brzeska, a Memoir.* New York: New Directions, 1970.

Ramachandran, V. S. Why do gentlemen prefer blondes? *Medical Hypotheses, 48,* 1997, 19–20.

Rebuffe-Scrive, M. Regional adipose tissue metabolism in men and in women during menstrual cycle, pregnancy, lactation, and menopause. *International Journal of Obesity, 11,* 1987, 347–355.

———, Culberg, G., Lundberg, P. A., Lindstedt, G. & Bjorntorp, P. Anthropometric variables and metabolism in polycystic ovarian disease. *Hormone Metabolic Research, 21,* 1989, 391–397.

Rhodes, G. *Superportraits: Caricatures and Recognition.* East Sussex, UK: Psychology Press, 1996.

———, Profitt, F., Grady, J. M. & Sumich, A. Facial symmetry and the perception of beauty. *Psychonomic Bulletin and Review,* in press.

——— & Tremewan, T. Averageness, exaggeration, and facial attractiveness. *Psychological Science, 7,* 1996, 105–110.

Ricketts, R. M. Divine proportions in facial aesthetics. *Clinics in Plastic Surgery, 9,* 1982, 401–422.

Ritter, J. M., Casey, R. J. & Langlois, J. H. Adults' responses to infants varying in appearance of age and attractiveness. *Child Development, 62,* 1991, 68–82.

Rodin, J. *Body Traps: How to Overcome Your Body Obsessions and Liberate the Real You.* London: Vermilion, 1992.

Romm, S. *The Changing Face of Beauty.* St. Louis: Mosby, 1992.

Rooks, N. M. *Hair Raising: Beauty, Culture, and African American Women.* New Brunswick: Rutgers University Press, 1996.

Rosa, R. A. & Kotkin, H. C. That acquired masseteric look. *Journal of Dentistry for Children,* March-April 1996, 105–107.

Rose, P. *Parallel Lives: Five Victorian Marriages.* New York: Vintage, 1984.

Rosenberg, A. & Kagan, J. Iris pigmentation and behavioral inhibition. *Developmental Psychobiology, 20,* 1987, 377–392.

——— & Kagan, J. Physical and physiological correlates of behavioral inhibition. *Developmental Psychobiology, 22,* 1989, 753–770.

Rowley, H. A., Baluja, S. & Kanade, T. Human face detection in visual scenes. *Carnegie Mellon Computer Science Technical Report,* CMU-CS-95, 158R, November 1995.

Rozin, P. & Fallon, A. Body image, attitudes to weight, and misperceptions of figure preferences of the opposite sex: A comparison of men and women in two generations. *Journal of Abnormal Psychology, 97,* 1988, 342–345.

RuPaul. *Lettin It All Hang Out: An Autobiography.* New York: Hyperion, 1995.

Russell, B. *The Conquest of Happiness.* New York: Liveright, 1958.

Russell, K., Wilson, M. & Hall, R. *The Color Complex: The Politics of Skin Color Among African Americans.* New York: Anchor, 1993.

Sadalla, E. K., Kenrick, D. T., Vershure, B. Dominance and heterosexual attraction. *Journal of Personality and Social Psychology, 52,* 1987, 730–738.

Salvia, J., Sheare, J. B. & Algozzine, B. Facial attractiveness and personal-social development. *Journal of Abnormal Child Psychology, 3,* 1973, 171–178.

Samuels, C. A., Butterworth, G., Roberts, T., Graupner, L. & Hole, G. Facial aesthetics: Babies prefer attractiveness to symmetry. *Perception, 23,* 1994, 823–831.

Santayana, G. *The Sense of Beauty: Being the Outline of Aesthetic Theory.* New York: Dover, 1955.

Sapolsky, R. The importance of a well-groomed child. *Science, 277,* 1997, 1620–1621.

Scafidi, F. A., Field, T. & Schanberg, S. M. Factors that predict which preterm infants benefit most from massage therapy. *Journal of Development and Behavioral Pediatrics, 14,* 1993, 176–180.

Schanberg, S. M. & Field, T. M. Sensory deprivation stress and supplemental stimulation in the rat pup and preterm human neonate. *Child Development, 58,* 1987, 1431–1447.

Schenkenberg, M. *New Rules.* New York: Universe/Rizzoli, 1997.

Schjeldahl, P. "Notes on Beauty." In *Uncontrollable Beauty: Toward a New Aesthetic.* Ed. B. Beckley with D. Shapiro. New York: Allworth Press, 1998.

Schneider, S. K. *Vital Mummies: Performance Design for the Show-window Mannequin.* New Haven: Yale University Press, 1995.

Schor, J. B. *The Overspent American: Upscaling, Downshifting, and the New Consumer.* New York: Basic Books, 1998.

Schwartz, L. *The Computer Artist's Handbook.* New York: W. W. Norton, 1992.

Scutt, D. & Manning, J. T. Symmetry and ovulation. *Human Reproduction, 11,* 1996, 2477–2480.

Sebesta, J. L. & Bonfonte, L. *The World of Roman Costume.* Madison: University of Wisconsin Press, 1994.

Selmi, P. "Fashions and Luxuries in the Political Mentality of the Venetian Republic." In *I Mestieri Della Moda A Venezia Serenissima: The Arts of Fashion in Venice from the 13th to the 18th Century.* Curator D. D. Poli. Catalogue to exhibiton at the Equitable Gallery, New York City, 1995.

Semler, D. E. Some aspects of adaptation in a polymorphism for breeding colours in the Threespine stickleback (Gasterosteus aculeatus). *Journal of Zoology, 165,* 1971, 291–302.

Sergios, P. & Cody, J. Importance of physical attractiveness and social assertiveness skills in male homosexual dating behavior and partner selection. *Journal of Homosexuality, 12,* 1985, 71–84.

Seymour, M. *Ottoline Morrell: Life on the Grand Scale.* London: Sceptre, 1993.

Sforza, L. L. & Sforza, F. *The Great Human Diasporas: The History of Diversity and Evolution.* Reading, MA: Addison-Wesley, 1996.

Short, R. V. Testes weight, body weight, and breeding systems in primates, *Nature, 293,* 1981, 55.

Siever, M. D. Sexual orientation and gender as factors in socioculturally acquired vulnerability to body dissatisfaction and eating disorders. *Journal of Consulting and Clinical Psychology, 62,* 1994, 252–260.

Sigall, H. & Aronson, E. Liking for an evaluator as a function of her physical attractiveness and nature of the evaluations. *Journal of Experimental Social Psychology, 5,* 1969, 93–100.

————— & Landy, D. Radiating beauty: Effects of having a physically attractive partner on person perception. *Journal of Personality and Social Psychology, 28,* 1973, 218–224.

————— & Ostrove, N. Beautiful but dangerous: Effects of offender attractiveness and nature of the crime on juridic judgment. *Journal of Personality and Social Psychology, 31,* 1975, 410–414.

Signorile, M. *Life Outside. The Signorile Report on Gay Men: Sex, Drugs, Muscles, and the Passages of Life.* New York: HarperCollins, 1997.

Singh, D. Adaptive significance of female physical attractiveness: Role of waist-to-hip ratio. *Journal of Personality and Social Psychology, 65,* 1993, 293–307.

_____. Body fat distribution and perception of desirable body shape by young black men and women. *International Journal of Eating Disorders, 16,* 1994, 289–294.

_____. Female judgment of male attractiveness and desirability for relationships: role of waist-to-hip ratio and financial status. *Journal of Personality and Social Psychology, 69,* 1995, 1089.

_____ & Luis, S. Ethnic and gender consensus for the effect of waist-to-hip ratio on judgment of women's attractiveness. *Human Nature, 6,* 1995, 51–65.

_____ & Young, R. K. Body weight, waist-to-hip ratio, breasts, and hips: Role in judgments of female attractiveness and desirability for relationships. *Ethology and Sociobiology, 16,* 1995, 483–507.

Skrebneski, V. *The Art of Haute Couture.* Text by Laura Jacobs. New York: Abbeville Press, 1995.

Snyder, M., Tanke, E. D. & Berscheid, E. Social perception and interpersonal behavior: On the self-fulfilling nature of social stereotypes. *Journal of Personality and Social Psychology, 35,* 1977, 656–666.

Sobal, J. & Stunkard, A. J. Socioeconomic status and obesity: A review of the literature. *Psychological Bulletin, 105,* 1989, 260–275.

Sontag, S. "The double standard of aging." In *Psychology of Women.* Ed. J. Williams. New York: Academic Press, 1979.

Sparling, J. Penile erections: shape, angle, and length. *Journal of Sex and Marital Therapy, 23,* 1997, 195–207.

Sroufe, R, Chaiken, A., Cook, R. & Freeman, V. The effects of physical attractiveness on honesty: A socially desirable response. *Personality and Social Psychology Bulletin, 3,* 1977, 59–62.

Stass, W. & Willis, F. N., Jr. Eye contact, pupil dilation, and personal preference. *Psychonomic Science, 7,* 1967, 375–376.

Steinbeck, J. *The Wayward Bus.* New York: Viking, 1947.

Steinberg, L. *The Sexuality of Christ in Renaissance Art and in Modern Oblivion.* New York: Pantheon, 1983.

Steinberger, E., Rodriguez-Rigau, L., Smith, K. D., Held, N. The menstrual cycle and plasma testosterone levels in women with acne. *Journal of the American Academy of Dermatology, 4,* 1981, 54–58.

Steinem, G. *Revolution from Within: A Book of Self-Esteem.* Boston: Little Brown, 1992.

Stern, K. & McClintock, M. K. Regulation of ovulation by human pheromones. *Nature, 392,* 1998, 177–179.

Stewart, D. E. Reproductive functioning in eating disorders. *Annals of Medicine, 24,* 1992, 287–291.

_____, Robinson, E., Goldbloom, D. S. & Wright, C. Infertility and eating disorders. *American Journal of Obstetrics and Gynecology, 163,* 1990, 1196–1199.

Stoddard, J. T. Composite photography. *Century, 33,* 1887, 750–757.

Stoddardt, D. M. *The Scented Ape: The Biology and Culture of Human Odor.* Cambridge: Cambridge University Press, 1991.

Stoller, R. J. *Presentations of Gender.* New Haven: Yale University Press, 1985.

Strathern, A. "Dress, Decoration, and Art in New Guinea." In *Man as Art.* Photographs by Malcolm Kirk. San Francisco: Chronicle Books, 1993.

Strathern, M. The self in self-decoration. *Oceania, 48,* 1979, 241–257.

Strong, G. R. *The Portraits of Elizabeth I.* London: Thames and Hudson, 1987.

Stubbs, R. H. Penis lengthening: A retrospective review of 300 consecutive cases. *Canadian Journal of Plastic Surgery, 5,* 1997, 93–100.

Sweeney, M. M. Remarriage of men and women: The role of socioeconomic prospects. *Journal of Family Issues, 18,* 1997, 479–502.

_____. Women, men, and changing families: The shifting economic foundation of marriage. *Center for Demography and Ecology Working Paper No. 97-14.* Madison, Wisconsin, 1997.

Symons, D. *The Evolution of Human Sexuality.* New York: Oxford University Press, 1979.

————. "Beauty is in the adaptations of the beholder: The evolutionary psychology of human female sexual attractiveness." In *Sexual Nature, Sexual Culture.* Ed. P. R. Abramson & S. D. Pinkerton. Chicago: University of Chicago Press, 1995.

Synott, A. "Truth and goodness, mirrors and masks—Part I: A sociology of beauty and the face." *British Journal of Sociology, 40,* 1988, 607–636.

Tahara Y., Tsunawake, N., Yukawa, K., Yamaski, M., Nishiyama, K., Urata, H., Katsuno, K. & Fukuyama, Y. Sex differences in interrelationships between percent body fat (%fat) and waist-to-hip ratio in healthy male and female adults. *Annals of Physiological Anthropology, 13,* 1994, 293–301.

Tanke, E. D. Dimensions of the physical attractiveness stereotype: A factor/analytic study. *Journal of Psychology, 110,* 1982, 63–74.

Taylor, P. A. & Glenn, N. D. The utility of education and attractiveness for female status attainment through marriage. *American Sociological Review, 41,* 1976, 484–498.

Terry, R. L. Effects of facial transformations on accuracy of recognition. *Journal of Social Psychology, 134,* 1993, 483–492.

Thakerer, J. N. & Iwawaki, S. Cross-cultural comparisons in interpersonal attraction of females toward males. *Journal of Social Psychology, 108,* 1979, 121.

Thornhill, R. Female preference for the pheromone of males with low fluctuating asymmetry in the Japanese scorpionfly (Panorpa japonica: mecoptera). *Behavioral Ecology, 3,* 1992, 277–283.

———— & Gangestad, S. W. Human facial beauty: Averageness, symmetry, and parasite resistance. *Human Nature, 4,* 1993, 237–269.

———— & Gangestad, S. W. Human fluctuating asymmetry and sexual behavior. *Psychological Science, 5,* 1994, 297–302.

————, Gangestad, S. W. & Comer, R. Human female orgasm and mate fluctuating asymmetry. *Animal Behavior. 50,* 1995, 1601–1615.

———— & Moller, A. P. Developmental stability, disease, and medicine. *Biological Reviews of the Cambridge Philosophical Society, 72,* 1997, 497–548.

———— & Sauer, K. P. Genetic sire effects on the fighting ability of sons and daughters and mating success of sons in the scorpionfly (Panorpa vulgaris). *Animal Behavior, 43,* 1992, 255–264.

Tichet, J., Vol, S., Balkau, B., Le Clesiau, H. & D'Hour, A. Android fat distribution by age and sex: The waist hip ratio. *Diabetes Metabolism, 19,* 1993, 273–276.

Tolstoy, L. *Childhood, Boyhood, Youth.* Trans. Michael Scammel. New York: McGraw-Hill, 1964.

Tooby, J. & Cosmides, "Introduction: Evolutionary Psychology and Conceptual Integration." In Barkow, Cosmides & Tooby, 1992, 3–15.

———— & Cosmides, "The psychological foundations of culture." In Barkow, Cosmides, & Tooby, 1992, 19–136.

———— & Cosmides, L. "Evolutionary Psychology: A Primer," unpublished ms. University of California, Santa Barbara,

Tovee, M. J., Mason, S. M., Emery, J. L., McClusky, S. E. & Cohen-Tovee, E. M. Supermodels: stick insects or hourglasses? *Lancet, 350,* 1997, 1474–1475.

Townsend, J. M. Mate selection criteria: A pilot study. *Ethology and Sociobiology, 10,* 1989, 241–253.

———— & Levy, G. D. Effect of potential partners' physical attractiveness and socioeconomic status on sexuality and partner selection. *Journal of Sexual Behavior, 19,* 1990, 149–164.

_____ & Levy, G. D. Effects of potential partners' costume and physical attractiveness on sexuality and partner selection. *Journal of Psychology, 124,* 1990, 371–389.

Trasko, M. *Daring Do's: A History of Extraordinary Hair.* Paris: Flammarion, 1994.

Udry, J. R. Structural correlates of feminine beauty preferences in Britain and the United States: A comparison. *Sociology and Social Research, 49,* 1965, 330–342.

_____ & Eckland, B. K. Benefits of being attractive: Differential payoffs for men and women. *Psychological Reports, 54,* 1984, 47–56.

Vale, V. & Juno, A. *Modern Primitives: An Investigation of Contemporary Adornment and Ritual.* San Francisco: Re/Search Publications, 1989.

Valéry, P. "Some simple reflections on the body." In *Fragments for a History of the Human Body.* Part 2. Ed. Michel Feher with Ramona Naddaff and Nadia Tazi. New York: Zone, 1989.

Van den Berghe, P. L. & Frost, P. Skin color preference, sexual dimorphism and sexual selection: A case of gene culture co-evolution? *Ethnic and Racial Studies, 9,* 1986, 87–113.

Veblen, T. *The Theory of the Leisure Class.* New York: Penguin, 1994.

Voloshinov, A. V. Symmetry as a superprinciple of science and art. *Leonardo, 29,* 1996, 109–113.

Vroon, P. *Smell: The Secret Seducer.* New York: Farrar, Straus, and Giroux, 1994.

de Waal, F. *Chimpanzee Politics: Power and Sex Among Apes* (London: Cape, 1982).

_____. *Good Natured: The Origins of Right and Wrong in Humans and Other Animals.* Cambridge: Harvard University Press, 1996.

Wagatsuma, E. & Kleinke, C. L. Ratings of facial beauty by Asian-American and Caucasian females, *Journal of Social Psychology, 109,* 1979, 299–300.

Wagatsuma, H. The social perception of skin color in Japan. *Daedalus, 96,* 1967, 407–443.

Walster, E., Aronson, V., Abrahams, D. & Rottman, L. Importance of physical attractiveness in dating behavior. *Journal of Personality and Social Psychology, 4,* 1966, 508–516.

Warner, M. *From the Beast to the Blonde: On Fairy Tales and Their Tellers.* New York: Noonday (Farrar, Straus, and Giroux), 1995.

Wass, P., Waldenstrom, U., Rossner, S. & Hellberg, D. An android body fat distribution in females impairs the pregnancy rate of in-vitro fertilization-embryo transfer. *Human Reproduction, 12,* 1997, 2057–2060.

Watson, J. B. *Behaviorism.* New York: W. W. Norton, 1925.

Webster, M. & Driskell, J. E. Beauty as status. *American Journal of Sociology, 89,* 1983, 140–165.

Wedekind, C., Furi, S. Body odour preferences in men and women: Do they aim for specific MHC combinations or simple heterozygosity? *Proceedings of the Royal Society, B264,* 1997, 1471–1479.

_____, Seebeck, T., Bettens, F. & Paepke, A. J. MHC-dependent mate preferences in humans. *Proceedings of the Royal Society of London, B260,* 1995, 245–249.

Weller, A. Communication through body odor. *Nature, 392,* 1998, 126–127.

Westermarck, E. *The History of Human Marriage.* London: Macmillan, 1921.

Whalen, P. J., Rauch S. L., Etcoff, N. L., McInerny, S. C., Lee, M. B., Jenike, M. A. Masked presentations of emotional facial expressions modulate amygdala activity without explicit knowledge. *Journal of Neuroscience, 18,* 1998, 411–418.

Wiederman, M. W. & Allgeier, E. R. Gender differences in mate selection criteria: Sociobiological or socioeconomic explanation? *Ethology and Sociobiology, 13,* 1992, 115–124.

Wilkinson, G. S., Presgraves, D. C. & Crymes, L. Male eye span in stalk-eyed flies indicates genetic quality by meiotic drive suppression. *Nature, 391,* 1998, 276.

Williams, T. *A Streetcar Named Desire.* New York: Signet, 1974.

Wilson, P. R. Perceptual distortion of height as a function of ascribed academic status. *Journal of Social Psychology, 74,* 1968, 97–102.

Winer, J. "The Floating Lightbulb." In *Minding the Body: Women Writers on Body and Soul.* Ed. P. Foster. New York: Anchor, 1994.

Wolf, N. *The Beauty Myth: How Images of Beauty Are Used Against Women.* New York: Anchor, 1992.

Wolf, W. C. & MacDonald, M. The earnings of males and marital disruption. Center for Demography and Ecology Working Paper, 78-14. Madison Wisconsin, 1978.

⸺ & MacDonald, M. The earnings of men and remarriage. *Demography, 16,* 1979, 389–399.

Wolfe, T. "Funky Chic." In *20 Years of Rolling Stone: What a Long, Strange Trip It's Been.* Ed. Jann Wenner. New York: Friendly Press, 1987.

Wolff, P. "Observations on the Early Development of Smiling." In *Determinants of Infant Development.* Vol 2. Ed. B. Foss. New York: Wiley, 1963.

Wong, R. C. & Ellis, C. N. Physiologic skin changes in pregnancy. *Journal of the American Academy of Dermatology, 10,* 1984, 929–943.

Yalom, M. *A History of the Breast.* New York: Alfred Knopf, 1997.

Yamaguchi, M. K. & Oda, M. Measuring and creating different facial images for age and gender. *ATR Technical Report,* 1996.

Zaadstra, B. M., Seidell, J. C., Van Noord, P. A. H., te Velde, E. R., Habbema, J. D. F., Vrieswjik, B., Karbaat, J. Fat and female fecundity: Prospective study of effect of body fat distribution on conception rates. *British Medical Journal, 306,* 1993, 484–487.

Zahavi, A. Mate selection: A selection for a handicap. *Journal of Theoretical Biology, 53,* 1975, 205–214.

Zajonc, R. B., Adelman, P. K., Murphy, S. T. & Niedenthal, P. M. Convergence in the physical appearance of spouses. *Motivation and Emotion, 11,* 1987, 335–346.

Zebrowitz, L. A. *Reading Faces: Window to the Soul.* New York: Westview, 1997.

⸺ & Montepare, J. M. Impressions of babyfaced individuals across the life span. *Developmental Psychology, 28,* 1992, 1143–1152.

⸺, Montepare, J. M. & Lee, H. K. They don't all look alike: Individuated impressions of other racial groups. *Journal of Personality and Social Psychology, 65,* 1993, 85–101.

Zentner, M. R. and Kagan, J. Perception of music by infants. *Nature, 383,* 1996, 29.

Zuckerman, M. & Driver, R. E. What sounds beautiful is good: The vocal attractiveness stereotype. *Journal of Nonverbal Behavior, 13,* 1989, 67–82.

⸺, Miyake, K. & Hodgins, H. S. Cross-channel effects of vocal and physical attractiveness and their implications for interpersonal perception. *Journal of Personality and Social Psychology, 60,* 1991, 545–554.

Zuk, M. "Parasites and Bright Birds: New Data and a New Prediction." In *Bird-Parasite Interactions: Ecology, Evolution and Behavior.* Ed. J. E. Loye and M. Zuk. Oxford: Oxford University Press, 1982.

Photos and Illustrations

Page 1. Hans Sebald Beham, "A Man's Head; A Woman's Head," 1542. Courtesy of the Fogg Art Museum, Harvard University Art Museums, Gray Collection of Engravings Fund. Photography by David Mathews. © President and Fellows of Harvard College, Harvard University.

Page 29. The Perception Laboratory, University of St. Andrews, "Composite Portraits," 1998. In the nineteenth century, Sir Francis Galton created composite portraits by aligning the eyes on photographic negatives. The top row of this figure shows three (of thirty) faces that were blended in this way to create the composite at top right. Recent advances in computer technology allow the average positions of facial features to be calculated for a set of images. Warping each face to this average shape (bottom row) before blending allows crisp composites to be generated (bottom right). Photo courtesy of the Perception Laboratory, University of St. Andrews, Fife, Scotland. © University of Saint Andrews.

Page 55. Sir Francis Galton, "Bridges and Bridegrooms: Newspapers and Photographs," 1904–1905. Courtesy of University College, London. Papers of Sir Francis Galton (Galton 162/J). Photography of Robert Masters.

Page 89. Giovanni Marco Pitteri, "Carlo Goldoni," 1754. Courtesy of the Fogg Art Museum, Harvard University Art Museums, bequest of Henry George Berg Fund, "Gift in Honor of John Coolidge." Photograph by Harvard Photographic Services. © President and Fellows of Harvard College, Harvard University.

Page 131. The Perception Laboratory, University of St. Andrews, "Masculinised and Feminised Female Face Shapes," 1998. A composite of thirty Japanese women's faces (center); masculinized 30 percent (left); and feminised 30 percent (right). Viewers tend to prefer the feminine face shape to the average or masculinized shape. Courtesy of the Perception Laboratory, University of St. Andrews, Fife, Scotland. © University of St. Andrews.

Page 167. Dr. Devendra Singh, "Michelangelo's David with 1.0, .9, and .7 Waist-to-Hip Ratio." Courtesy of Dr. Devendra Singh, Department of Psychology, University of Texas at Austin. © Dr. Devendra Singh.

Page 205. Webb Chappell, "The Lingua Trekka." One of the smart clothes ensembles modeled at the Beauty and the Bits Fashion Show, MIT, October 15, 1997. The lingua trekka includes an ambient linguistic device with speakers for simultaneous language translation at the back neckline. The chest piece is equipped with a removable mini-screen and keyboard. The midriff tattoo is a universal immunization, protecting the wearer against pathogens. Designers: Nao Muramatsu, Hisayoshi Kuroda, Junko Ito, and Keiko Minomo. Technology Collaborators: Sumit Basu, Jennifer Healy, and Thad Starner. © Webb Chappell.

Page 231. Andre Kertesz, "New York City," 1979. © Estate of Andre Kertesz.

Acknowledgments

I am indebted to the many people who helped me in so many ways during the years I researched and wrote (and brooded over) this book. My colleagues shared their discoveries and insights with me. I thank Don Symons for years of illuminating discussions on beauty. Anne Becker, Hans Breiter, Helena Cronin, Paul Ekman, Victor Johnston, Sandy Pentland, David Perrett, Steve Pinker, Gill Rhodes, Devendra Singh, and Randy Thornhill all provided valuable information and insights. I thank Paul Ekman for being a brilliant and inspiring mentor and true friend. Lauren Cooper was an assistant extraordinaire, and Steven Antonik, Greta Buck, and Anne Grossetete did heroic work ferreting out even the most obscure references. I also thank Pat Claffey, Catherine Carter, Caroline Kerrigan, and Krista Tibbs for their assistance.

I am grateful for the support of my home institution, Harvard Medical School and the Massachusetts General Hospital, Department of Psychiatry. I owe particular thanks to Hans Breiter, Bruce Rosen, Michael Jenike and Ned Cassem for their support. I was generously supported in my research by the Lynne M. Reid Fellowship. Sandy Pentland and the members of his Vision and Modeling group at the MIT Media Lab have been valued colleagues and collaborators. I thank Baback Moghaddam in particular for his help in setting up many a demo. I also thank Dr. Shigeru Akamatsu for inviting me to the ATR Laboratories in Kyoto, Japan, and I thank my gracious host Masami Yamaguchi for sharing her innovative work with me.

Stan Sclaroff, Raffaella Rumiati, Claudio Luzzatti, Danielle Barry, Mark Halman, Trevor Darrell, Michael Jordan, Jamie Hamilton, Cindy Bogatka, Kathleen Peets, Lauren Cooper, William Green, Rhya Fisher, Janet Cahn, and Philip Greenspun sustained me with their friendship, advice, affection, confidence, and perhaps most of all, with their sense of humor. John Magee was my ingenious partner in research at MIT, and has been there for me in all the important ways. Kevin Oregan and I had wonderful talks about fashion, and model Hoyt Richards and speech therapist Sam Chwat graciously answered my many questions during interviews. Geoffrey Cowley's astute questions helped me to sharpen my ideas. My editor Betsy Lerner provided careful guidance, insight, and the title! I also thank Matt Ellis for his much valued assistance. Jeremy Taylor, Ian Penton-Voak, David Perrett, Robert Marsters, Robert Klein, Marsit Erb, and most of all Robert Gurbo helped me put together the extraordinary set of visual images for this book. Finally I thank Simson Garfinkel and Beth Rosenberg for loaning me their wonderful home in Martha's Vineyard for three months of writing.

I reserve a special thank you for John Brockman and Katinka Matson. This book would not have happened without their encouragement, their support, and their vision. I cannot thank my mother enough for all the love and support she has given me. My sister Linda has been my friend and confidante. Perhaps more than anyone, she understands how this book came to be. I thank her and Chuck for their encouragement, and for many discussions of art and beauty during their dazzling dinners. My dog Max sat witness to every word on these pages and was the perfect canine companion. I think often about my father, who would have enjoyed seeing this book, and who was its inspiration.

Index

315

around eyes and brow, 111, 114–15, 146, 152
breast implants, 5, 189–90, 226
at earlier ages, 112
evolution of changes in, 146–47
gender differences in use of, 60
for lips, 152, 153
men's, 82, 111, 125, 178, 181–82
mental health and, 19–20
on noses, 111, 146, 152
proportional measurement systems for, 17, 140
rates of, 5, 19, 82, 110–11
secrecy lifted from, 223–24
Cosmides, Leda, 20, 22, 23, 45
Couturiers, 220–22, 227–28
See also Fashion
Crawford, Cindy, 12, 58, 66, 224, 237
Criminals, physical appearance of, 49, 144
Cultural relativism, 21, 135–36
Culture
perceptions of beauty influenced by, 22–23, 31, 34
social science model emphasizing, 20–22
transition phase in, 80–83
Cunningham, Michael, 139, 156, 158

Darwin, Charles
on beauty through diversity, 5
on blushing, 107
on courting displays by birds, 169
on cultural relativity of beauty, 135–36
influence of physiognomy on, 22
on romance and reproduction, 69–70
tattooing noted by, 99
Dating. *See* Romance
Deformities, as sign of evil, 41
Democracy, envy as basis of, 68–69
Depression, 8, 20, 240
Designer labels, 222–23, 227
Diamond, Jared, 73, 116
Dieting. *See* Eating disorders; Weight
Dior, Christian, 195, 208, 221
Discrimination. *See* Beauty prejudice (lookism); Racism
Divorce, 76, 81, 82, 103–4
Dress. *See* Clothing; Fashion; Fashion industry

Drugs
for depression, 20
for preventing hair loss, 124–25
residue in hair from, 123
Dürer, Albrecht, 11, 16, 17, 19, 78, 149

Eating disorders, 124, 193
gender differences in, 60, 202–3
reproduction and, 203–4
responsibility for, 26, 201
Education, appearance and grades, 48
Egypt (ancient), 96–97, 102
Ekman, Paul, 22–23, 115
Eliot, George, 244–45
Elizabeth I, Queen (of England), 13, 101, 103–4, 214
Emotions
facial expression of, 22–23, 32, 113–15, 135, 159, 160–61, 162, 164
happiness, 48, 85–88
infants arousing, in adults, 34–38
love, 14, 50–51, 107
England
fashion and status in, 214–15, 217, 219, 220
obesity in, 196, 199, 200–1
penalties for cosmetic use, 102–3
See also Elizabeth I, Queen (of England)
Envy, of women, 67–69
Estrogen, 111–12, 113, 153, 176, 191
Europe
birth of fashion in, 213–14
cosmetics industry in, 95–96
obesity in, 200–1
standard of beauty from, 43, 117–18
sumptuary laws in, 218–19
See also England; France
Evangelista, Linda, 12, 224, 237
Evolution
average proportions fostered by, 145–46, 150
beauty as biological adaptation, 14, 22–24, 233–34
facial expressions/features, 113, 134
health in populations susceptible to disease, 59–60
of nude hairless body, 92–94
relationship between appearance and safety, 40

Homeliness. *See* Ugliness
Homosexuality
  eating disorders and, 202–3
  importance of appearance and, 59,
    62, 63–64
  pornography market and, 61
Honesty, impacts of beauty on, 44–45
Hormone replacement therapy, 112
Hormones
  facial features and, 152, 153
  hair loss and, 124, 125
  height and, 176
  immune functions and, 171
  odors and, 238–40
  skin and, 111–13
  weight and, 191

Income
  importance to women of, 75–76,
    77–78, 79, 80
  weight and, 200
  women's wage-earning ability, 66,
    79–82, 173
  See also Careers/professions; Status
Infants
  birth weights of, 145, 199
  emotional responses to, 34–38
  facial features of, in adults, 154–55,
    156, 158
  familial resemblance patterns, 38–
    39
  importance of skin-to-skin touching,
    95
  mothers' reactions to, 35–36, 37–38
  recognition of beauty by, 31–32,
    163
  responses to faces, 31–34
  sense of smell, 241
  skin and hair colors, 104–5
Intelligence
  beauty as indication of, 43, 48
  clothing reflecting, 228–29
  importance of, in marriage partners,
    66

Japan
  female vocal pitch, 237
  shape of eyes, 135
  shy posing in, 154
  skin color preferences in, 106
  as style center, 226, 228
  sumptuary laws in, 18
  use of cosmetics in, 96, 101

Jesus Christ
  artistic representations of, 209
  physical beauty of, 19
  teachings on beauty, 18
Johnston, Victor, 151, 153, 163
Jones, Douglas, 106, 119, 138–39,
  151

Kagan, Jerome, 32, 127–28
Keats, John, 123, 233
Kissinger, Henry, 78, 176
Koinophilia. *See* Average appearance

Languages, universal grammar
    underlying, 22
Lavater, Johann Caspar, 22, 43
Leonardo da Vinci, 16, 140, 149
Lindzey, Gardner, 20, 21, 22
Lips. *See* Faces; Mouths
Lipstick, 102, 107, 153
Lookism, 25, 83
Love
  blushing and, 107
  nude photographs and, 51
  quest for beauty to gain, 14
LPCs (late positive component of
    event-related potentials), 163

Makeup. *See* Cosmetics
Male dominance
  affecting mating preferences, 80,
    158
  beauty and, 74–75, 155–56, 159
Males. *See* Biology; Bodies, human;
    Men
Marriage
  age differences within, 64
  appearance and prospects for, 65–
    67
  earning capabilities and, 75–76,
    80–81
  hair styles and, 120–21
  income differences in, 79–81
  physical similarities of partners,
    147, 175
  between races, 119
  women working and, 81–82
  See also Romance
Mathematical measurement systems.
    See Proportion
Mead, Margaret, 21, 95
Media
  ages of film stars, 64–65